Part of the **MASTERING ESSENTIAL MATH SKILLS** Series

— New 2nd Edition —

Whole Numbers and Integers

20 minutes a day to success

Richard W. Fisher

Math **+ −**
Essentials **X ÷**
MARINA, CALIFORNIA

*Mastering Essential Math Skills: **Whole Numbers and Integers, 2nd Edition***
Copyright © 2021, Richard W. Fisher.
Printed and bound in the United States of America.

For information, please contact Math Essentials, 469 Logan Way, Marina, CA 93933.

Although the author and publisher have made every effort to ensure
the accuracy and completeness of information contained in this book, we assume
no responsibility for errors, inaccuracies, omissions, or any inconsistency herein.
Any slighting of people, places, or organizations is unintentional.

First printing 2021
ISBN 978-1-7372633-0-2

Notes to Parents, Teachers, and Students

What sets this book apart from other books is its approach. It is not just a math book, but a system of teaching math. Each daily lesson contains three key parts: **Review Exercises**, **Helpful Hints**, and **Problem Solving**. Teachers have flexibility in introducing new topics, but the book provides them with the necessary structure and guidance. The teacher can rest assured that essential math skills in this book are being systematically learned.

This easy-to-follow program requires only fifteen or twenty minutes of instruction per day. Each lesson is concise and self-contained. The daily exercises help students to not only master math skills, but also maintain and reinforce those skills through consistent review—something that is missing in most math programs. Skills learned in this book apply to all areas of the curriculum, and consistent review is built into each daily lesson. Teachers and parents will also be pleased to note that the lessons are quite easy to correct.

This book is based on a system of teaching that was developed by a math instructor over a thirty-year period. This system has produced dramatic results for students. The program quickly motivates students and creates confidence and excitement that leads naturally to success.

Please read the "How to Use This Book" section on the next page and let this program help you to produce dramatic results with your math students.

Tips for Using the Online Video Tutorials

- Go to www.mathessentials.net
- Click on the Videos button.
- There are 3 sources for videos: Book 1, Book 2, and Pre-algebra.
- Passwords are in red print.
- The video menus are set up in chapters.
- Find the chapter that contains the topic that you want. For example, the chapter might be FRACTIONS. "Adding Fractions With Unlike Denominators" might be the video that you would like to select.
- To get the best learning experience when viewing the video, work right along with the instructor using a paper and pencil.
- For more advanced Algebra videos go to www.nononsensealgebra.com and click on "Video library" in the upper right-hand corner. A complete video library will appear.

After you've explored the menus and used the videos a few times,
using the Online Video Tutorials will become easy.

How to Use This Book

This book is best used on a daily basis. The first lesson should be carefully gone over with students to introduce them to the program and familiarize them with the format. It is hoped that the program will help your students to develop an enthusiasm and passion for math that will stay with them throughout their education. As you go through these lessons every day, you will soon begin to see growth in the student's confidence, enthusiasm, and skill level. The students will maintain their mastery through the daily review.

Step 1: The students are to complete the review exercises, showing all their work. After completing the problems, it is important for the teacher or parent to go over this section with the students to ensure understanding.

Step 2: Next comes the new material. Use the "Helpful Hints" section to help introduce the new material. Be sure to point out that it is often helpful to come back to this section as the students work independently. This section often has examples that are very helpful to the students.

Step 3: It is highly important for the teacher to work through the two sample problems with the students before they begin to work independently. Working these problems together will ensure that the students understand the topic, and prevent a lot of unnecessary frustration. The two sample problems will get the students off to a good start and will instill confidence as the students begin to work independently.

Step 4: Each lesson has problem solving as the last section of the page. It is recommended that the teacher go through this section, discussing key words and phrases, and also possible strategies. Problem solving is neglected in many math programs, and just a little work each day can produce dramatic results.

Step 5: Solutions are located in the back of the book. Teachers may correct the exercises if they wish, or have the students correct the work themselves.

Table of Contents

Whole Numbers and Integers

Whole Numbers

Integers

Math Resource Center

Review Exercises

Note to the students and teachers: This section will include review problems from all the topics covered in this book. Here are some simple problems with which to get started.

1. $\begin{array}{r} 314 \\ + \ 43 \\ \hline \end{array}$

3. $\begin{array}{r} 7 \\ 2 \\ + \ 4 \\ \hline \end{array}$

5. $7 + 4 + 5 =$

2. $\begin{array}{r} 603 \\ + \ 24 \\ \hline \end{array}$

4. $\begin{array}{r} 426 \\ + \ 313 \\ \hline \end{array}$

6. $\begin{array}{r} 22 \\ 16 \\ + \ 11 \\ \hline \end{array}$

Helpful Hints	1. Line up numbers on the right side. 2. Add the ones first. 3. Remember to regroup when necessary. 4. Place commas in the answer when necessary.	* "Sum" means to total or add.

* Samples S1 and S2 may be worked together by teacher and students.

S1. $\begin{array}{r} 243 \\ 62 \\ + \ 514 \\ \hline \end{array}$

S2. $\begin{array}{r} 345 \\ 423 \\ + \ 165 \\ \hline \end{array}$

1. $\begin{array}{r} 42 \\ 56 \\ + \ 15 \\ \hline \end{array}$

2. $\begin{array}{r} 516 \\ 54 \\ + \ 213 \\ \hline \end{array}$

3. $\begin{array}{r} 517 \\ 326 \\ 143 \\ + \ 14 \\ \hline \end{array}$

4. $\begin{array}{r} 813 \\ 236 \\ 17 \\ + \ 6 \\ \hline \end{array}$

5. $\begin{array}{r} 543 \\ 227 \\ 312 \\ + \ 415 \\ \hline \end{array}$

6. $\begin{array}{r} 5 \\ 23 \\ 416 \\ + \ 243 \\ \hline \end{array}$

7. $415 + 436 + 317 =$

8. $23 + 34 + 16 + 47 =$

9. Find the sum of 443, 514, and 716.

10. Find the sum of 127, 23, 16, and 246.

1.
2.
3.
4.
5.
6.
7.
8.
9.
10.
Score

Problem Solving	There are 42 students in the fifth grade, 56 students in the sixth grade, and 36 students in the seventh grade. How many students are there altogether?

Review Exercises

1. 426
    + 23

2. 526
    42
    + 163

3. Find the sum of 23, 34, and 26.

4. 342
    426
    + 325

5. 427
    316
    12
    + 423

6. 36
    324
    573
    + 11

| **Helpful Hints** | Use what you have learned to solve each problem.
REMEMBER: 1. Line up numbers on the right side.
                2. Add the ones first.
                3. Regroup when necessary   * "Sum" means add.
                4. Place commas in the answer when necessary. |

* Samples S1 and S2 may be worked together by teacher and students.

S1. 426
    343
    + 516

S2. 16
    247
    346
    + 125

1. 702
    337
    + 225

2. 23
    45
    + 67

3. 627
    423
    504
    + 213

4. 6
    17
    289
    + 363

5. 724
    427
    395
    + 367

6. 600
    528
    396
    + 27

7. 7 + 27 + 48 + 247 =

8. 763 + 29 + 372 + 16 =

9. Find the sum of 96, 73, 44, and 75.

10. Find the sum of 213, 426, 516, and 423.

1.

2.

3.

4.

5.

6.

7.

8.

9.

10.

Score

| **Problem Solving** | Monique earned $52 on Monday. On Tuesday, Wednesday, and Thursday she earned $56 each day. How much did she earn altogether? |

Review Exercises

1.   36
    47
    16
 + 92

2.   727
    423
 +   75

3.   824
    216
    724
 + 316

4.   967 + 843 + 96 =

5.   72 + 16 + 49 =

6.   Find the sum of 14, 15, 16, and 17.

Helpful Hints	When writing large numbers, place commas every three numerals, starting from the right. This makes them easier to read.	**Example:** 21 million, 234 thousand, 416 21,234,416

S1.   4,236
    147
 + 3,236

S2.   7,217
    685
 + 36,125

1.   4,736
    2,136
 + 3,523

2.   17,236
    3,175
 +   4,297

3.   1,312,516
 + 2,316,475

4.   7,137,236
    353,246
 +   7,506

5.   23,726
    17,532
    6,573
 + 23,247

6.   72,506
    1,343
    15,208
 +   5,123

7.   Find the sum of 1,342, 1,793, and 4,562.

8.   78,623 + 4,579 + 22,396 =

9.   72,214 + 16,738 + 29,143 + 17 =

10.   103,246 + 724,516 + 72,173 =

1.

2.

3.

4.

5.

6.

7.

8.

9.

10.

Score

Problem Solving	One country has a population of 7,143,600. Another country has a population of 16,450,395. A third country has a population of 12,690,000. What is the total population of all three countries?

Review Exercises

1.   376
      493
   +   16

2.   23,463 + 7,764 + 21,976 =

3.   72,176
      43,267
   + 19,237

4.   27,967,204
   + 5,137,264

5.   95,623
       7,429
      19,234
   + 89,342

6.   Find the sum of
     3,426,497 and
     7,136,095.

Helpful Hints	Use what you have learned to add the following problems. Practice reading your answers.	**Example:** 23 million, 106 thousand, 749
		23,106,749

S1.   76,142
        7,913
    + 16,408

S2.   313,426
     2,462,417
   + 3,526,008

1.   27,167
      35,225
   + 16,342

2.   1,234,617
      3,426,124
   +   4,316,243

3.   7,163
      2,476
   + 3,276

4.   17,226
      26,319
      15,314
   + 27,976

5.   16
        326
      7,764
   + 12,316

6.   76,174
      133,367
      243,776
   + 712,777

7.   76,724 + 55,726 + 5,723 + 74,126 =

8.   7,134,243 + 3,712,050 + 3,516,312 =

9.   Find the sum of 42,916, 47,993, and 6,716.

10.   7 + 17 + 278 + 42,563 =

1.
2.
3.
4.
5.
6.
7.
8.
9.
10.
Score

Problem Solving	A school district has three high schools. One school has 2,463 students, another school has 3,985 students, and the third school has 1,596 students. What is the total number of high school students in the district.

Review Exercises

1. 33
    45
    + 76

2. 7,716
    12,727
    + 23,496

3. 7,124,563
    5,236,907
    + 7,369,704

4. 776
    397
    + 442

5. 753,476 + 569,736 + 7,842 + 39,673 =

6. Find the sum of 72, 96, 72, 9, and 88.

Helpful Hints	1. Line up the numbers on the right side. 2. Subtract the ones first. 3. Regroup when necessary. 4. It may be necessary to regroup more than once. 5. "Find the difference" and "how much more" means to subtract.	**Examples:** $\overset{8}{7}\overset{}{\cancel{9}}3$ -75 = 718 $\overset{5\ 11}{\cancel{6}\cancel{2}3}$ -254 = 369

S1.   567
    - 183

S2.   4,352
    - 2,171

1.   339
    - 24

2.   623
    - 52

3.   6,153
    - 758

4.   5,231
    - 2,456

5.   1,387
    - 739

6.   7,383
    - 2,285

7. Find the difference between 763 and 47.

8. Subtract 527 from 3,916.

9. 7,249 - 779 =

10. 789 is how much more than 298?

1. _____

2. _____

3. _____

4. _____

5. _____

6. _____

7. _____

8. _____

9. _____

10. _____

Score

Problem Solving	846 students attend Vargas School and 943 students attend Hoover School. How many more students attend Hoover School than Vargas School?

Review Exercises

1. 726
    849
  + 308

3.   7,614
    - 1,557

5.   5,613
    - 752

2.   721
    - 443

4. Find the sum of 455, 673, and 998.

6. 32,456,175
     3,214,915
  + 6,318,425

| **Helpful Hints** | 1. Use what you have learned to solve the following problems.
2. It may be necessary to regroup more than once. | "find the difference"
"is how much more"
"is how much less" | All Mean Subtraction |

S1.   856
    - 297

S2.   9,613
    - 2,247

1.   715
   - 242

2.   324
   - 149

3.   3,137
   - 1,252

4.   6,334
   - 2,175

5.   7,123
   - 2,456

6.   7,156
   - 877

7. 27,346 - 15,472 =

8. Subtract 797 from 2,314.

9. 127 is how much less than 2,496?

10. Find the difference between 7,692 and 4,764.

1.	
2.	
3.	
4.	
5.	
6.	
7.	
8.	
9.	
10.	
Score	

| **Problem Solving** | Ahmir earned 86 dollars on Friday. He earned 28 dollars less than this amount on Saturday. How much did he earn on Saturday? |

Review Exercises

1. $\begin{array}{r} 395 \\ 428 \\ + \ 376 \\ \hline \end{array}$

3. $5,123 - 1,672 =$

5. $\begin{array}{r} 376 \\ 19 \\ 426 \\ + \ 442 \\ \hline \end{array}$

2. $7,162 - 396 =$

4. Find the difference between 8,961 and 279.

6. $3,152 - 496 =$

Helpful Hints

1. Line up the numbers on the right side.
2. Subtract the ones first.
3. It may be necessary to regroup more than once.

Examples:

$\begin{array}{r} {}^{6}\ {}^{10}\ {}^{9}{}^{1} \\ \cancel{7,1}\cancel{0}3 \\ - \ \ \ 677 \\ \hline 6,426 \end{array}$ $\begin{array}{r} {}^{5}\ {}^{9}\ {}^{9}{}^{1} \\ \cancel{6,0}\cancel{0}\cancel{0} \\ - \ 1,634 \\ \hline 4,366 \end{array}$

S1. $\begin{array}{r} 601 \\ - \ 356 \\ \hline \end{array}$

S2. $\begin{array}{r} 700 \\ - \ 267 \\ \hline \end{array}$

1. $\begin{array}{r} 90 \\ - \ 67 \\ \hline \end{array}$

2. $\begin{array}{r} 705 \\ - \ 176 \\ \hline \end{array}$

3. $\begin{array}{r} 4,012 \\ - \ 1,345 \\ \hline \end{array}$

4. $\begin{array}{r} 700 \\ - \ 238 \\ \hline \end{array}$

5. $\begin{array}{r} 8,000 \\ - \ 758 \\ \hline \end{array}$

6. $\begin{array}{r} 5,207 \\ - \ 1,539 \\ \hline \end{array}$

7. Find the difference between 13,012 and 10,796.

8. Subtract 7,689 from 9,026.

9. $70,023 - 63,345 =$

10. What number is 5,021 less than 12,506?

#	
1.	
2.	
3.	
4.	
5.	
6.	
7.	
8.	
9.	
10.	
Score	

Problem Solving

A theatre has 1,150 seats. If 859 of them are taken, how many seats are empty?

Review Exercises

1. $\begin{array}{r} 500 \\ -\ 276 \\ \hline \end{array}$

2. $\begin{array}{r} 3,196 \\ 742 \\ +\ \ 276 \\ \hline \end{array}$

3. Find the sum of 72, 95, 63, and 47.

4. $\begin{array}{r} 5,012 \\ -\ \ 763 \\ \hline \end{array}$

5. $\begin{array}{r} 7,126 \\ -\ 3,147 \\ \hline \end{array}$

6. $5,000 - 768 =$

| **Helpful Hints** | Use what you have learned to solve the following problems. | **Examples:** $\begin{array}{r} {}^{4}\ {}^{11}\ {}^{9}{}^{1} \\ \cancel{5,2\cancel{0}1} \\ -\quad 475 \\ \hline 4,726 \end{array}$ $\begin{array}{r} {}^{2}\ {}^{9}\ {}^{9} \\ \cancel{3,00}{}^{1}0 \\ -\ 1,235 \\ \hline 1,765 \end{array}$ |

S1. $\begin{array}{r} 6,002 \\ -\ 1,456 \\ \hline \end{array}$

S2. $\begin{array}{r} 7,000 \\ -\ \ 967 \\ \hline \end{array}$

1. $\begin{array}{r} 900 \\ -\ \ 77 \\ \hline \end{array}$

2. $\begin{array}{r} 5,004 \\ -\ 2,576 \\ \hline \end{array}$

3. $\begin{array}{r} 2,010 \\ -\ 1,346 \\ \hline \end{array}$

4. $\begin{array}{r} 5,600 \\ -\ \ 797 \\ \hline \end{array}$

5. $\begin{array}{r} 3,102 \\ -\ 1,453 \\ \hline \end{array}$

6. $\begin{array}{r} 50,000 \\ -\ \ 7,542 \\ \hline \end{array}$

7. $71,005 - 2,709 =$

8. Subtract 2,511 from 7,005.

9. What number is 57 less than 500?

10. $76,500 - 29,308 =$

1.	
2.	
3.	
4.	
5.	
6.	
7.	
8.	
9.	
10.	
Score	

| **Problem Solving** | A company earned 95,000 dollars this year. Last year the company earned 78,500 dollars. How much more did the company earn this year than last year? |

Review Exercises

1. 375
    427
    + 28

2. 32,176
    3,724
    + 15,756

3. Find the sum of 24, 296, 752, and 897.

4. 7,121
    - 2,430

5. 7,001
    - 2,765

6. 4,000
    - 2,173

Helpful Hints	Use what you have learned to solve the following problems.

S1.   562
      7,124
     15,765
    + 1,456

S2. 4,001
    - 1,327

1. 516
     39
    + 77

2. 714
    - 253

3. 5,317
     967
    + 6,765

4. 7,000
    - 4,789

5. 6,102
    - 1,799

6.   29
     37
     46
    + 53

7. 23,076 - 13,097 =

8. Find the sum of 197, 368, and 427.

9. How much more is 921 than 143?

10. 72,176 + 7,996 + 72,199 =

1.	
2.	
3.	
4.	
5.	
6.	
7.	
8.	
9.	
10.	
Score	

Problem Solving	Pedro wants to buy a bike that costs $850. If he has saved $755, how much more does he need to be able to buy the bike?

Review Exercises

1. 70
   + 26

2. 76,175 - 2,963 =

3. 325 + 72 + 765 + 89 =

6. 72,172
   5,347
   56
   + 2,347

4. 72,105
   - 7,367

5. Find the difference of
   7,723 and 10,197.

Helpful Hints	1. Use what you have learned to solve the following problems. 2. Line up numbers on the right. 3. Be careful when regrouping.

S1. 7,000
    - 1,345

S2. 37,673
    7,742
    + 7,396

1. 800
   - 124

2. 7,001
   - 1,237

3. 27
   347
   2,436
   + 7,998

4. 35,021
   - 7,130

5. 32 + 96 + 33 + 37 =

6. 27
   99
   97
   68
   + 53

7. 976 + 279 + 342 + 196 =

8. 70,001 - 2,617 =

9. How much more is 7,526 than 3,017?

10. Find the sum of 22,463, 7,296, and 7,287.

1.
2.
3.
4.
5.
6.
7.
8.
9.
10.

Score

Problem Solving	71,503 people visited the museum on Monday. 80,106 people visited the museum on Friday. How many more people visited the museum on Friday than on Monday?

Review Exercises

1. 501
    - 267

2. 334
    79
    617
    + 22

3. 6,002 - 3,667 =

4. 27 + 33 + 36 + 52 =

5. Find the difference of 8,012 and 796.

6. 27,763
    4,245
    7,342
    + 767

| Helpful Hints | 1. Line up numbers on the right.
2. Multiply the ones first.
3. Regroup when necessary.
4. "Product" means to multiply. | **Examples:** $\overset{1\ 1}{644}$
 x   3
 1,932 | $6,\overset{3\ 2}{076}$
 x   4
 24,304 |

S1. 526
     x   3

S2. 3,254
     x   6

1.   67
     x   3

2.   74
     x   6

3.   427
     x   6

4.   4,214
     x   6

5.   3,056
     x   8

6.   7,256
     x   6

7. 7,036 x 7 =

8. 4 x 3,849 =

9. Find the product of 8,762 and 6.

10. Multiply 9 and 3,872.

1.	
2.	
3.	
4.	
5.	
6.	
7.	
8.	
9.	
10.	
Score	

Problem Solving	If there are 365 days in each year, how many days are there in 7 years?

Review Exercises

1. 343
    x   4

2. 3,064
    x   7

3. 3,601 - 798 =

4. 6,000
    - 1,235

5. 375
    429
     63
    + 7

6. 763 x 4 =

Helpful Hints	Use what you have learned to solve the following problems.	***Remember** Put commas in the answer if necessary and practice reading the answers.

S1. 7,234
     x   6

S2. 9,007
     x   8

1. 728
    x 6

2. 9,012
    x   9

3. 23,726
    x   7

4. 7,632
    x   6

5. 9,730
    x   9

6. 7,009
    x   4

7. Find the product of 8 and 2,643.

8. 8,012 x 7 =

9. 6 x 23,728 =

10. 12,429 x 6 =

1.

2.

3.

4.

5.

6.

7.

8.

9.

10.

Score

Problem Solving	Ty needs a total of 450 points to win a prize. If she has already earned 298 points, how many more points does she need to win the prize?

Review Exercises

1. 725
    x   6

2. 3,012
    x   9

3. 727 + 33 + 526 + 724 =

4. 23,102 is how much more than 7,256?

5. 7,001
    - 3,674

6. 375
    47
    462
    + 578

Helpful Hints	1. Line up numbers on the right. 2. Multiply the ones first. 3. Multiply the tens second. 4. Add the two products. 5. Place commas in the product if necessary.

Examples:

```
      43            437
x     32       x     26
      86            2622
+   1290       +    8740
   1,376          11,362
```

S1. 43
     x 24

S2. 246
     x   53

1. 82
    x 53

2. 46
    x 17

3. 85
    x 47

4. 436
    x   25

5. 336
    x   50

6. 706
    x   47

7. Find the product of 16 and 37.

8. 92 x 47 =

9. 763 x 45 =

10. 460 x 33 =

1.
2.
3.
4.
5.
6.
7.
8.
9.
10.
Score

Problem Solving	A school has 25 classrooms. If each classroom has 35 desks, how many desks are there altogether?

Review Exercises

1. 3,015
    - 176

2. 3,145
    x   7

3.   48
    x 36

4.   423
    x   25

5. 7,163
    5,427
    + 3,136

6. 7,562 - 1,999 =

Use what you have learned to solve the following problems.
* Remember to put commas in your answer.

S1.   402
    x   26

S2.   356
    x   44

1.   26
    x 50

2.   56
    x 23

1. _____

2. _____

3.   55
    x 46

4.   216
    x   64

5.   248
    x   76

6.   476
    x   75

3. _____

4. _____

5. _____

6. _____

7. 92 x 103 =

7. _____

8. 468 x 26 =

8. _____

9. Find the product of 65 and 70.

9. _____

10. _____

10. Multiply 608 and 73.

Score _____

3,216 students attend Washington Middle School. If 2,016 of the students are boys, how many girls attend the school?

Review Exercises

1. $\begin{array}{r} 42 \\ \times\ 36 \\ \hline \end{array}$

2. $\begin{array}{r} 407 \\ \times\ 28 \\ \hline \end{array}$

3. $\begin{array}{r} 400 \\ \times\ 35 \\ \hline \end{array}$

4. $\begin{array}{r} 7,612 \\ -\ 1,357 \\ \hline \end{array}$

5. $36 + 37 + 19 + 62 =$

6. $7,000 - 2,836 =$

Helpful Hints	1. Line up numbers on the right. 2. Multiply the ones first. 3. Multiply the tens second. 4. Multiply the hundreds last. 5. Add the products. 6. Put commas in the product.	**Examples:**	$\begin{array}{r} 243 \\ \times\ \ 336 \\ \hline 1458 \\ 7290 \\ +\ 72900 \\ \hline 81,648 \end{array}$	$\begin{array}{r} 673 \\ \times\ \ 307 \\ \hline 4711 \\ 0000 \\ +\ 201900 \\ \hline 206,611 \end{array}$

S1. $\begin{array}{r} 153 \\ \times\ 423 \\ \hline \end{array}$

S2. $\begin{array}{r} 724 \\ \times\ 526 \\ \hline \end{array}$

1. $\begin{array}{r} 247 \\ \times\ 315 \\ \hline \end{array}$

2. $\begin{array}{r} 246 \\ \times\ 137 \\ \hline \end{array}$

3. $\begin{array}{r} 244 \\ \times\ 302 \\ \hline \end{array}$

4. $\begin{array}{r} 364 \\ \times\ 503 \\ \hline \end{array}$

5. $\begin{array}{r} 543 \\ \times\ 414 \\ \hline \end{array}$

6. $\begin{array}{r} 269 \\ \times\ 400 \\ \hline \end{array}$

7. $521 \times 337 =$

8. Find the product of 208 and 326.

9. Multiply 500 and 822.

10. $444 \times 366 =$

1.
2.
3.
4.
5.
6.
7.
8.
9.
10.
Score

Problem Solving

A factory can produce 3,050 cars per week.
How many cars can the factory produce in one year?
(Hint: There are 52 weeks in a year.)

Review Exercises

1. 7,632
    558
    3,627
    +   36

2. 3,000
    - 2,176

3. 2,172
    x   6

4.   46
    x 23

5.   263
    x   30

6.   526
    x 704

Helpful Hints	Use what you have learned to solve the following problems. 1. Put commas in the answers. 2. Practice reading the answers.

S1. 444 x 225	S2. 350 x 906	1. 906 x 717	2. 500 x 743	1. _____
				2. _____
				3. _____
				4. _____
3. 323 x 435	4. 648 x 755	5. 672 x 986	6. 246 x 932	5. _____
				6. _____

7. 219 x 406 =

8. 700 x 610 =

9. 763 x 908 =

10. Multiply 723 and 847.

7. _____

8. _____

9. _____

10. _____

Score

Problem Solving	Each bus can hold 112 students. How many students can 7 buses hold?

Review Exercises

1. 408
    x   6

2. Find the product
    of 24 and 36.

3. 7,665
    8,327
    + 9,342

4. 3,121
    - 2,344

5. 742
    x   6

6. 408
    x   27

| **Helpful Hints** | Use what you have learned to solve the following problems. |

S1. 426
     x   25

S2. 613
     x   425

1. 37
    x   5

2. 705
    x   6

3. 2,346
    x   7

4. 58
    x   72

5. 245
    x   63

6. 128
    x   547

7. Find the product of 15 and 726.

8. 700 x 657 =

9. 33 x 610 =

10. 308 x 407 =

1.	
2.	
3.	
4.	
5.	
6.	
7.	
8.	
9.	
10.	
Score	

| **Problem Solving** | If there are 1,440 minutes in a day, how many minutes are there in a week? |

Review Exercises

1. $365 + 19 + 342 =$

2.
$$\begin{array}{r} 627 \\ \times\ \ 5 \\ \hline \end{array}$$

3.
$$\begin{array}{r} 7,136 \\ -\ \ \ 807 \\ \hline \end{array}$$

4.
$$\begin{array}{r} 214 \\ \times\ \ \ 7 \\ \hline \end{array}$$

5.
$$\begin{array}{r} 206 \\ \times\ \ \ 77 \\ \hline \end{array}$$

6.
$$\begin{array}{r} 364 \\ \times\ 500 \\ \hline \end{array}$$

| **Helpful Hints** | 1. Use what you have learned to solve the following problems.
2. Carefully line up the numbers.
3. Put commas in answers when necessary
4. Practice reading the answers. |

S1.
$$\begin{array}{r} 6,007 \\ \times\ \ \ \ 5 \\ \hline \end{array}$$

S2.
$$\begin{array}{r} 347 \\ \times\ \ \ 55 \\ \hline \end{array}$$

1.
$$\begin{array}{r} 404 \\ \times\ \ \ 5 \\ \hline \end{array}$$

2.
$$\begin{array}{r} 7,002 \\ \times\ \ \ \ 6 \\ \hline \end{array}$$

3.
$$\begin{array}{r} 8,916 \\ \times\ \ \ \ 7 \\ \hline \end{array}$$

4.
$$\begin{array}{r} 96 \\ \times\ \ 80 \\ \hline \end{array}$$

5.
$$\begin{array}{r} 718 \\ \times\ \ 63 \\ \hline \end{array}$$

6.
$$\begin{array}{r} 424 \\ \times\ 726 \\ \hline \end{array}$$

7. $3 \times 720 =$

8. $9 \times 7,856 =$

9. $67 \times 715 =$

10. $208 \times 763 =$

1.

2.

3.

4.

5.

6.

7.

8.

9.

10.

Score

| **Problem Solving** | 726 people attend the theatre on Saturday. On Sunday 827 people attended the theatre. How many more people attended the theatre on Sunday than on Saturday? |

Review Exercises

1. 624
    - 317

2. 347
    x   5

3. 72,196
    + 16,429

4. Find the difference between 7,912 and 995.

5. 136
    x   22

6. 200 x 309 =

Helpful Hints

1. Divide
2. Multiply
3. Subtract
4. Begin Again

Examples:

$$\begin{array}{r} 15\,r2 \\ 3\overline{)47} \\ -3\downarrow \\ \hline 17 \\ -15 \\ \hline 2 \end{array}$$

$$\begin{array}{r} 9\,r5 \\ 6\overline{)59} \\ -54 \\ \hline 5 \end{array}$$

Remember! The remainder must be less than the divisor.

S1. $4\overline{)17}$

S2. $5\overline{)49}$

1. $3\overline{)74}$

2. $8\overline{)47}$

3. $6\overline{)78}$

4. $5\overline{)93}$

5. $7\overline{)84}$

6. $5\overline{)79}$

7. 77 ÷ 8 =

8. 96 ÷ 4 =

9. $\dfrac{65}{4}$

10. $\dfrac{47}{2}$

1.	
2.	
3.	
4.	
5.	
6.	
7.	
8.	
9.	
10.	

Problem Solving

Pencils come in boxes of 72. If the pencils are divided equally among 4 students, how many pencils will each student receive?

Score

Review Exercises

1. $2\overline{)17}$

2. $5\overline{)89}$

3. $7\overline{)45}$

4. $205 - 99 =$

5. $\begin{array}{r} 165 \\ \times 7 \\ \hline \end{array}$

6. $\begin{array}{r} 47 \\ \times 33 \\ \hline \end{array}$

Helpful Hints	Use what you have learned to solve the following problems. 1. Divide 2. Multiply 3. Subtract 4. Begin Again 5. REMEMBER! The remainder must be less than the divisor.

S1. $4\overline{)73}$ S2. $6\overline{)27}$ 1. $3\overline{)29}$ 2. $3\overline{)45}$

3. $8\overline{)93}$ 4. $2\overline{)97}$ 5. $7\overline{)60}$ 6. $5\overline{)29}$

7. $77 \div 5 =$ 8. $\dfrac{99}{4}$

9. $33 \div 4 =$ 10. $65 \div 5 =$

1.	
2.	
3.	
4.	
5.	
6.	
7.	
8.	
9.	
10.	
Score	

Problem Solving	Mr. Johnson's salary is $6,500 per month. How much does he earn per year? (Hint: How many months are in a year?)

Review Exercises

1. 6)‾38‾

2. 5)‾69‾

3. 7,103 - 2,617 =

4. 526 x 7 =

5. $\begin{array}{r} 164 \\ \times\ \ 23 \\ \hline \end{array}$

6. 73 + 16 + 57 + 76 =

Helpful Hints

1. Divide
2. Multiply
3. Subtract
4. Begin Again

Examples:

$$\begin{array}{r} 171\ r2 \\ 3\overline{)515} \\ -3\!\downarrow \\ \hline 21 \\ -21\!\downarrow \\ \hline 05 \\ -3 \\ \hline 2 \end{array}$$

$$\begin{array}{r} 203 \\ 4\overline{)812} \\ -8\!\downarrow \\ \hline 01 \\ -0\!\downarrow \\ \hline 12 \\ -12 \\ \hline 0 \end{array}$$

REMEMBER!
The remainder must be less than the divisor.

S1. 2)‾523‾   S2. 6)‾274‾   1. 3)‾972‾   2. 2)‾419‾

3. 7)‾933‾   4. 6)‾727‾   5. 8)‾916‾   6. 6)‾850‾

7. 3)‾936‾   8. 5)‾607‾   9. 3)‾776‾   10. 8)‾717‾

1.
2.
3.
4.
5.
6.
7.
8.
9.
10.

Problem Solving

A bakery produced 324 cookies. If 9 cookies are placed into each package, how many packages will the bakery need?

Score

Review Exercises

1. $3\overline{)57}$

2. $7\overline{)89}$

3. $5\overline{)616}$

4. $5\overline{)374}$

5. $7\overline{)924}$

6. $7\overline{)450}$

Helpful Hints	Use what you have learned to solve the following problems.

S1. $2\overline{)375}$ S2. $4\overline{)377}$ 1. $8\overline{)916}$ 2. $8\overline{)630}$

3. $7\overline{)679}$ 4. $8\overline{)976}$ 5. $4\overline{)508}$ 6. $6\overline{)248}$

7. $3\overline{)400}$ 8. $7\overline{)693}$ 9. $5\overline{)778}$ 10. $6\overline{)609}$

1.

2.

3.

4.

5.

6.

7.

8.

9.

10.

Score

Problem Solving	The Martinez family took a vacation and drove 450 miles each day for 12 days. How many miles did they drive during their vacation?

Review Exercises

1. 3$\overline{)69}$

2. 5$\overline{)667}$

3.
$$\begin{array}{r} 213 \\ \times\ \ 47 \end{array}$$

4.
$$\begin{array}{r} 7{,}710 \\ \times\ \ \ 697 \end{array}$$

5. 900 x 614 =

6.
$$\begin{array}{r} 712 \\ 14 \\ +\ \ \ 47 \end{array}$$

Helpful Hints

1. Divide
2. Multiply
3. Subtract
4. Begin Again

Examples:

$$\begin{array}{r} 1708 \\ 3\overline{)5124} \\ -3\!\downarrow\ \ \ \\ \overline{21}\ \ \ \ \\ -21\!\downarrow\ \ \\ \overline{02}\ \ \ \\ -\ 0\!\downarrow \\ \overline{24} \\ -24 \\ \overline{0} \end{array}$$

$$\begin{array}{r} 448\ r2 \\ 4\overline{)1794} \\ -16\!\downarrow\ \\ \overline{19}\ \ \\ -16\!\downarrow \\ \overline{34} \\ -32 \\ \overline{2} \end{array}$$

REMEMBER!
The remainder must be less than the divisor.

S1. 3$\overline{)9052}$

S2. 5$\overline{)3352}$

1. 2$\overline{)9235}$

2. 2$\overline{)7362}$

3. 6$\overline{)6821}$

4. 5$\overline{)2240}$

5. 5$\overline{)7666}$

6. 4$\overline{)7123}$

7. 5$\overline{)61{,}235}$

8. 2$\overline{)24{,}362}$

9. 4$\overline{)41{,}236}$

10. 5$\overline{)22{,}014}$

1.	
2.	
3.	
4.	
5.	
6.	
7.	
8.	
9.	
10.	
Score	

Problem Solving

5 students opened a business and earned $16,550. If they divided the money equally, how much would each student receive?

Review Exercises

1. $3\overline{)526}$ 2. $3\overline{)605}$ 3. $7\overline{)2634}$

4. $5\overline{)6175}$ 5. $6\overline{)13,252}$ 6. $7,100 - 768 =$

Helpful Hints	1. Use what you have learned to solve the following problems. 2. Remainders must be less than the divisor. 3. Zeroes may sometimes appear in the quotient.

S1. $6\overline{)3976}$ S2. $5\overline{)13,976}$ 1. $3\overline{)4123}$ 2. $5\overline{)1726}$

3. $8\overline{)8902}$ 4. $6\overline{)5324}$ 5. $6\overline{)9309}$ 6. $8\overline{)8016}$

7. $3\overline{)16,723}$ 8. $7\overline{)22,305}$ 9. $4\overline{)61,053}$ 10. $8\overline{)37,716}$

1.
2.
3.
4.
5.
6.
7.
8.
9.
10.
Score

Problem Solving	Yana took a 3-day hiking trip. The first day she hiked 13 miles, the second day she hiked 14 miles, and the third day she hiked 17 miles. How many miles did she hike altogether during the trip?

Review Exercises

1. $\begin{array}{r} 337 \\ 96 \\ + 349 \\ \hline \end{array}$

2. $\begin{array}{r} 32,105 \\ - 6,737 \\ \hline \end{array}$

3. $\begin{array}{r} 427 \\ \times 26 \\ \hline \end{array}$

4. $4\overline{)500}$

5. $7\overline{)1356}$

6. Find the sum of 39,764 and 79,743.

Helpful Hints	1. Divide 2. Multiply 3. Subtract 4. Begin again	* Remainders must be less than the divisor. * Zeroes may sometimes appear in the quotient.

S1. $3\overline{)901}$ S2. $7\overline{)7563}$ 1. $5\overline{)402}$ 2. $5\overline{)8500}$

3. $4\overline{)8413}$ 4. $8\overline{)3566}$ 5. $6\overline{)1526}$ 6. $5\overline{)6000}$

7. $4\overline{)8009}$ 8. $9\overline{)8765}$ 9. $7\overline{)6329}$ 10. $4\overline{)3217}$

1.

2.

3.

4.

5.

6.

7.

8.

9.

10.

Score

Problem Solving	A ream of paper contains 500 sheets. How many sheets of paper are there in 25 reams?

Review Exercises

1. Find the difference between 7,026 and 4,567.

2. $\begin{array}{r} 637 \\ \times\ 502 \\ \hline \end{array}$

3. $7\overline{)600}$

4. $5\overline{)5003}$

5. $2\overline{)3051}$

6. $7\overline{)1969}$

Helpful Hints

Use what you have learned to solve the following problems.

S1. $3\overline{)369}$

S2. $6\overline{)8772}$

1. $5\overline{)307}$

2. $9\overline{)157}$

3. $6\overline{)1213}$

4. $7\overline{)7012}$

5. $6\overline{)9736}$

6. $4\overline{)1398}$

7. $8\overline{)13,423}$

8. $5\overline{)14,387}$

9. $6\overline{)71,234}$

10. $5\overline{)15,355}$

1.
2.
3.
4.
5.
6.
7.
8.
9.
10.
Score

Problem Solving

A dairy produced 4,448 gallons of milk. If the milk is put into containers that hold 8 gallons each, how many containers will be needed?

Review Exercises

1. $6\overline{)718}$ 2. $2,007 - 865 =$ 3. $\begin{array}{r} 324 \\ \times\ 700 \\ \hline \end{array}$

4. $5\overline{)5007}$ 5. $6\overline{)1399}$ 6. $\begin{array}{r} 3,966 \\ 7,723 \\ +\ 4,564 \\ \hline \end{array}$

Helpful Hints	1. Divide 2. Multiply 3. Subtract 4. Begin again	Examples: $\begin{array}{r} 12\ r49 \\ 60\overline{)769} \\ -\ 60\downarrow \\ \hline 169 \\ -\ 120 \\ \hline 49 \end{array}$ $\begin{array}{r} 44\ r5 \\ 40\overline{)1765} \\ -\ 160\downarrow \\ \hline 165 \\ -\ 160 \\ \hline 5 \end{array}$

S1. $30\overline{)176}$ S2. $40\overline{)5732}$ 1. $70\overline{)238}$ 2. $50\overline{)376}$

3. $50\overline{)438}$ 4. $70\overline{)829}$ 5. $30\overline{)1682}$ 6. $20\overline{)1396}$

7. $40\overline{)8972}$ 8. $90\overline{)9096}$ 9. $40\overline{)3786}$ 10. $40\overline{)2396}$

1.
2.
3.
4.
5.
6.
7.
8.
9.
10.

Score

Problem Solving

A store puts eggs into boxes of a dozen. If there are 3,072 eggs, how many boxes will be needed?

Review Exercises

1. $2\overline{)69}$

2. $7\overline{)7019}$

3. $367 + 246 + 721 + 243 =$

4. $36 \times 304 =$

5. $6,108 - 976 =$

6. $3\overline{)6123}$

Helpful Hints	1. Use what you have learned to solve the following problems. 2. Zeroes may sometimes appear in the quotient. 3. Remember that remainders must be less than the divisor.

S1. $50\overline{)137}$ S2. $60\overline{)7612}$ 1. $20\overline{)396}$ 2. $20\overline{)187}$

3. $30\overline{)2312}$ 4. $30\overline{)7396}$ 5. $70\overline{)5724}$ 6. $70\overline{)9988}$

7. $50\overline{)9976}$ 8. $40\overline{)1766}$ 9. $80\overline{)2209}$ 10. $50\overline{)12,076}$

1.

2.

3.

4.

5.

6.

7.

8.

9.

10.

Score

Problem Solving	A school has 24 classes. Each class contains 32 students. How many students are there altogether in the school?

Review Exercises

1.
```
   463
    87
+  496
```

2.
```
  7,015
- 1,247
```

3. 963 - 713 =

4. 60)729

5. 70)367

6. 40)5276

| **Helpful Hints** | Sometimes it is easier to mentally round the divisor to the nearest power of ten. | Examples: $\begin{array}{r} 32\,r11 \\ 22\overline{)715} \\ -66\downarrow \\ \hline 55 \\ -44 \\ \hline 11 \end{array}$ | Think of: 20)715 |

S1. 32)673   S2. 32)279   1. 41)936   2. 28)617

3. 58)697   4. 83)762   5. 34)826   6. 31)976

7. 62)759   8. 53)609   9. 21)936   10. 42)866

1.
2.
3.
4.
5.
6.
7.
8.
9.
10.
Score

Problem Solving

A car traveled 448 miles. For each 32 miles it traveled, it consumed one gallon of gas. How many gallons of gas did it consume in traveling 448 miles?

Review Exercises

1. 326
    x 35

2. 324
    x 500

3. 27 + 89 + 76 + 14 =

5. 50$\overline{)826}$

6. 90$\overline{)7372}$

4. 3,000 - 756 =

Helpful Hints	Use what you have learned to solve the following problems. *Remember: Sometimes mentally rounding the divisor helps to make the problem easier.	Examples: $\begin{array}{r} 36 \\ 27\overline{)998} \\ -81 \\ \hline 188 \\ -162 \\ \hline 26 \end{array}$ Think of: $30\overline{)998}$

S1. 41$\overline{)912}$

S2. 51$\overline{)367}$

1. 52$\overline{)706}$

2. 39$\overline{)876}$

3. 47$\overline{)613}$

4. 22$\overline{)796}$

5. 12$\overline{)393}$

6. 25$\overline{)618}$

7. 49$\overline{)536}$

8. 22$\overline{)684}$

9. 28$\overline{)976}$

10. 42$\overline{)750}$

1.

2.

3.

4.

5.

6.

7.

8.

9.

10.

Score

Problem Solving

A theater had 12 rows of seats with 16 seats in a row.
If only 7 seats were empty, how many seats were occupied?

Review Exercises

1. $30\overline{)98}$ 2. $60\overline{)793}$ 3. $60\overline{)483}$

4. $32\overline{)426}$ 5. $32\overline{)275}$ 6. $28\overline{)405}$

Helpful Hints	Sometimes it is necessary to correct your estimate.	Examples: $$63\overline{)374} \atop -378 \text{ - too large}$$ with 6	$$\begin{array}{r} 5r59 \\ 63\overline{)374} \\ -315 \\ \hline 59 \end{array}$$

				1.
S1. $74\overline{)293}$	S2. $43\overline{)821}$	1. $38\overline{)197}$	2. $88\overline{)522}$	2.
				3.
				4.
3. $18\overline{)987}$	4. $22\overline{)178}$	5. $32\overline{)163}$	6. $14\overline{)886}$	5.
				6.
				7.
				8.
7. $34\overline{)649}$	8. $42\overline{)829}$	9. $36\overline{)721}$	10. $18\overline{)778}$	9.
				10.
				Score

Problem Solving	A machine can produce 75 parts in one hour. How many parts can it produce in 12 hours?

Review Exercises

1. 26
    x 3

2. 2,476
    x   7

3.   64
    x 23

4. 207
    x 25

5. 627
    x 400

6.   763
    x 509

Helpful Hints	Use what you have learned to solve the following problems. *Remember: Sometimes it is necessary to correct your estimate.

S1. 29⟌862   S2. 22⟌830   1. 27⟌851   2. 25⟌805

3. 19⟌572   4. 23⟌476   5. 31⟌927   6. 12⟌588

7. 18⟌787   8. 42⟌829   9. 14⟌886   10. 21⟌178

1.

2.

3.

4.

5.

6.

7.

8.

9.

10.

Score

Problem Solving	Children's tickets to a show are 5 dollars each and adult tickets are 7 dollars each. What would the total price be for 5 children's tickets and 4 adult tickets?

Review Exercises

1. 721
    - 356

2. 755 - 288 =

3. 60,123
    - 29,465

4. 779
    643
    247
    + 96

5. 637 + 975 +
    734 + 623 =

6. 7,256 is how much
    more than 5,909?

Helpful Hints

Use what you have learned to solve the following problems.

*Remember: Mentally round your divisor to the nearest power of ten.

Example:
$$32\overline{)6932} \quad 216\,r20$$
```
          216 r20
    32 ) 6932
        - 64 ↓
          53
        - 32 ↓
          212
        - 192
           20
```

Think of:
$$30\overline{)6932}$$

S1. $43\overline{)9250}$ S2. $32\overline{)6635}$ 1. $21\overline{)2645}$ 2. $51\overline{)7563}$

3. $42\overline{)9005}$ 4. $31\overline{)5126}$ 5. $27\overline{)8056}$ 6. $25\overline{)7651}$

7. $81\overline{)2654}$ 8. $22\overline{)1765}$ 9. $28\overline{)3976}$ 10. $12\overline{)9006}$

1.	
2.	
3.	
4.	
5.	
6.	
7.	
8.	
9.	
10.	
Score	

Problem Solving

A car traveled at the speed of 55 miles per hour for 8 hours. What was the total distance traveled?

Review Exercises

1. $3\overline{)106}$ 2. $7\overline{)6105}$ 3. $90\overline{)9672}$

4. $90\overline{)3807}$ 5. $42\overline{)506}$ 6. $42\overline{)369}$

Helpful Hints	Use what you have learned to solve the following problems.

S1. $34\overline{)1637}$ S2. $32\overline{)6795}$ 1. $41\overline{)3876}$ 2. $51\overline{)3962}$

3. $63\overline{)7196}$ 4. $81\overline{)7642}$ 5. $78\overline{)9053}$ 6. $58\overline{)3012}$

7. $25\overline{)7942}$ 8. $49\overline{)8006}$ 9. $12\overline{)4813}$ 10. $72\overline{)3099}$

1.	
2.	
3.	
4.	
5.	
6.	
7.	
8.	
9.	
10.	
Score	

Problem Solving	A farmer has 423 eggs. If he wants to put them into cartons that hold 36 eggs, how many full cartons will he have? How many eggs will be left over?

Review Exercises

1. 72,096
    7,423
    + 27,564

2. 7,505
    - 3,176

3. Find the sum of
    79,216 and 76,719.

4. Find the product of
    725 and 301.

5. Find the difference of
    6,000 and 890.

6. 336
    x 22

| **Helpful Hints** | Use what you have learned to solve the following problems. | *Mentally round 2-digit divisors when helpful. *Remainders must be less than the divisor. |

				1.
S1. $81\overline{)7716}$	S2. $28\overline{)1561}$	1. $2\overline{)77}$	2. $6\overline{)1989}$	2.
				3.
				4.
3. $5\overline{)1397}$	4. $40\overline{)647}$	5. $90\overline{)706}$	6. $90\overline{)3762}$	5.
				6.
				7.
				8.
7. $38\overline{)966}$	8. $23\overline{)278}$	9. $61\overline{)3344}$	10. $32\overline{)8096}$	9.
				10.
				Score

| **Problem Solving** | There were 96 students on a bus. At one stop 28 students got off. At the next stop 17 students got off. How many students were left on the bus? |

Review Exercises

1. $2\overline{)68}$ 2. $3\overline{)1964}$ 3. $30\overline{)279}$

4. $30\overline{)1367}$ 5. $62\overline{)707}$ 6. $62\overline{)4537}$

Helpful Hints	Use what you have learned to solve the following problems.

				1.
S1. $61\overline{)5872}$	S2. $27\overline{)2096}$	1. $5\overline{)97}$	2. $8\overline{)2605}$	2.
				3.
				4.
				5.
3. $9\overline{)7002}$	4. $60\overline{)967}$	5. $60\overline{)396}$	6. $80\overline{)6776}$	6.
				7.
				8.
7. $76\overline{)396}$	8. $76\overline{)962}$	9. $76\overline{)1796}$	10. $76\overline{)9908}$	9.
				10.
				Score

Problem Solving	Susan works 40 hours per week. If she is paid 12 dollars per hour, how much does she earn in 2 weeks?

Review of All Whole Number Operations

1. 456
    27
    + 626

2. 819
    746
    696
    + 638

3. 7,964 + 895 + 72,528 =

4. 7,096 + 2,716 + 779 + 84 =

5. 7,009 + 868 + 19 + 578 =

6. 742
    - 397

7. 6,291
    - 3,476

8. 9,051 - 2,766 =

9. 8,000 - 3,988 =

10. 9,051 - 2,766 =

11. 87
    x 3

12. 6,542
    x   7

13. 49
    x 63

14. 809
    x 76

15. 409
    x 278

16. 3$\overline{)425}$

17. 6$\overline{)1697}$

18. 40$\overline{)867}$

19. 53$\overline{)7976}$

20. 38$\overline{)1698}$

1.	
2.	
3.	
4.	
5.	
6.	
7.	
8.	
9.	
10.	
11.	
12.	
13.	
14.	
15.	
16.	
17.	
18.	
19.	
20.	

Final Review of All Whole Number Operations

1.   767
     455
     396
   + 228

2.   72,367
      9,768
   + 7,709

3. 16,221 + 6,872 + 9,796 + 15 =

4. 7 + 77 + 777 + 7,777 =

5. 9,697 + 7,636 + 8,964 =

6.   700
   - 265

7.   9,001
   - 786

8. 52,153 - 7,654 =

9. 5,000 - 2,357 =

10. 72,591 - 16,784 =

11.   308
    x   7

12.   7,010
    x   9

13.   87
    x 29

14.   978
    x   39

15.   963
    x 801

16. 7⟌412

17. 6⟌1392

18. 40⟌239

19. 63⟌8974

20. 29⟌17,236

1.	
2.	
3.	
4.	
5.	
6.	
7.	
8.	
9.	
10.	
11.	
12.	
13.	
14.	
15.	
16.	
17.	
18.	
19.	
20.	

Review Exercises

1. $775 + 639 + 426 =$

2. $\begin{array}{r} 7,163 \\ -269 \\ \hline \end{array}$

3. Find the difference between 7,455 and 3,672.

4. $7\overline{)1637}$

5. $70\overline{)367}$

6. $40\overline{)719}$

Helpful Hints

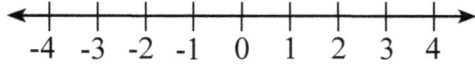
$-4\ -3\ -2\ -1\ \ 0\ \ 1\ \ 2\ \ 3\ \ 4$

Integers to the left of zero are negative and less than zero. Integers to the right of zero are positive and greater than zero. When two integers are on a number line, the one farthest to the right is greater.

Hint: Always find the sign of the answer first when working these problems.

Examples: The sum of two negatives is a negative.

$-7 + -5 = -$
(The sign is negative.)
$\begin{array}{r} 7 \\ +\ 5 \\ \hline 12 \end{array} = \boxed{-12}$

When adding a negative and a positive, the sign is the same as the integer farthest from zero, then subtract.

$-7 + 9 = +$
(The sign is positive.)
$\begin{array}{r} 9 \\ -\ 7 \\ \hline 2 \end{array} = \boxed{+2}$

S1. $-7 + 13 =$

S2. $-16 + -7 =$

1. $-17 + 26 =$

2. $-14 + -4 =$

3. $56 + -72 =$

4. $-14 + -13 =$

5. $-9 + 16 =$

6. $-47 + 85 =$

7. $-102 + -78 =$

8. $65 + -94 =$

9. $-23 + -47 =$

10. $65 + -85 =$

1.

2.

3.

4.

5.

6.

7.

8.

9.

10.

Problem Solving

There are 320 students going on a field trip. If each bus holds 55 students, how many buses will be needed for the field trip?

Score

Review Exercises

1. -32 + 18 =   2. 17 + -12 =   3. -32 + -46 =

4.   453   5. 32 $\overline{)409}$ 6. 32 $\overline{)197}$
   x   327

Helpful Hints	Use what you have learned to solve the following problems. *Remember to find the sign of the answer first.

S1. -96 + 105 =	S2. -99 + -86 =	1. -67 + 58 =	1.
			2.
			3.
2. -235 + -701 =	3. -56 + 19 =	4. -95 + -46 =	4.
			5.
			6.
5. -163 + 200 =	6. -423 + 208 =	7. -525 + -376 =	7.
			8.
			9.
8. 924 + -1,023 =	9. -346 + -295 =	10. 650 + -496 =	10.
			Score

Problem Solving	Maria bought a CD that cost 19 dollars. If she paid with a 50 dollar bill, how much change will she receive?

Review Exercises

1.   3,009
    x   7

2.   767
        96
     + 394

3.   72,052
    - 13,654

4. -75 + 60 =

5. -92 + -96 =

6. -105 + 142 =

| **Helpful Hints** | When adding more than two integers, group the negatives and positives separately, then add. | **Examples:**
-6 + 4 + -5 =
-11 + 4 = -
(Sign is negative.)
$\begin{array}{r} 11 \\ -\ 4 \\ \hline 7 \end{array}$ = -7 | 7 + -3 + -8 + 6 =
-11 + 13 = +
(Sign is positive.)
$\begin{array}{r} 13 \\ -\ 11 \\ \hline 2 \end{array}$ = +2 |

S1. -4 + 7 + -5 =

S2. -7 + 6 + -8 + 4 =

1. -7 + -5 + 15 =

2. 9 + -4 + -8 =

3. -16 + 22 + -12 =

4. 8 + -6 + 4 + 7 =

5. -22 + 40 + -24 =

6. -12 + 8 + -10 + 6 =

7. -12 + -7 + -14 =

8. -7 + 8 + -3 + -7 =

9. -40 + 18 + -16 + 12 =

10. -84 + 30 + -35 =

1.

2.

3.

4.

5.

6.

7.

8.

9.

10.

Score

| **Problem Solving** | A student received test scores of 84, 90, and 93. What was his average score? |

Review Exercises

1. $40\overline{)967}$

2. $40\overline{)384}$

3. $32\overline{)697}$

4. $32\overline{)1987}$

5. $-4 + 8 + -9 + 7 =$

6. $-35 + 40 + -77 =$

Helpful Hints	Use what you have learned to solve the following problems. Remember to group the negatives and positives separately, then add.

S1. $-7 + 9 + -8 =$

S2. $-5 + 12 + -13 - 15 =$

1. $-6 + -8 + 7 =$

2. $15 + -7 + -30 =$

3. $-50 + 17 + -22 =$

4. $16 + -12 + 18 + -8 =$

5. $-52 + 32 + -14 =$

6. $-20 + 14 + -12 + 32 =$

7. $62 + -101 + 15 =$

8. $-35 + -36 + -37 =$

9. $-12 + 21 + -16 + 40 =$

10. $-92 + 58 + -23 =$

1.

2.

3.

4.

5.

6.

7.

8.

9.

10.

Score

Problem Solving	A business needs 600 postcards to mail to customers. If postcards come in packages of 25, how many packages does the business need to buy?

Review Exercises

1. -17 + 5 + -16 =

2. -5 + -6 + 7 + 12 =

3. -14 + 25 + -32 + 6 =

4. 36 + 72 + 14 +
    96 + 23 =

5. 2,015 - 786 =

6.   786
     x   22

Helpful Hints

* To subtract integers means to add to its opposite. **Examples:**

-3 - -8 =   8
-3 + 8 = +   - 3
(Sign is positive.)   5 = +5

8 - 10 =   10
8 + -10 = -   - 8
(Sign is negative.)   -2 = -2

6 - -7 =   7
6 + 7 = +   + 6
(Sign is positive.)   13 = +13

S1. -7 - 6 =

S2. -7 - -8 =

1. 3 - -12 =

2. 16 - 19 =

3. -14 - -22 =

4. -17 - 14 =

5. 50 - -16 =

6. 48 - 14 =

7. -8 - 12 =

8. -72 - -54 =

9. -39 - 54

10. -63 - -94 =

1.	
2.	
3.	
4.	
5.	
6.	
7.	
8.	
9.	
10.	
Score	

Problem Solving

At night the temperature was 44 degrees. By morning it had dropped 52 degrees. What was the temperature in the morning?

Review Exercises

1. -7 + 6 + -8 =   2. 16 + -22 + -10 + 6 =   3. 32 + -46 + -16 =

4. -35 - 7 =   5. 16 - -4 =   6. 32 - 48 =

Helpful Hints	Use what you have learned to solve the following problems. Remember, to subtract means to add its opposite.
	Examples: -8 - 20 = 8 + -20 -8 - -20 = -8 + 20

S1. -12 - 16 =	S2. 50 - -62 =	1. -22 - -60 =	1.
			2.
			3.
2. 15 - 23 =	3. -24 - -36 =	4. -14 - 32 =	4.
			5.
			6.
5. 55 - -16 =	6. -39 - 40 =	7. -6 - 5 - 3 =	7.
			8.
			9.
8. -102 - -150 =	9. -220 - 214 =	10. -58 - -42 =	10.
			Score

Problem Solving	Together, Juan and James have earned 520 dollars. If Juan has earned 325 dollars, how much has James earned?

Review Exercises

1. -7 - 9 =

2. -12 - -10 =

3. 18 - -22 =

4. 16 + -7 + 12 =

5. -8 + 6 + -7 + 5 =

6. -22 + -72 + -13 =

Helpful Hints

Use what you have learned to solve the problems on this page. **Examples:**

-7 + 4 + -3 + 2 =   10
-10 + 6 = -   - 6
(Sign is negative.)   ――――
                        4 = (-4)

-7 - -6 =   7
-7 + 6 = -   - 6
(Sign is negative.) ――――
                    1 = (-1)

15 - 36 =   36
15 + -36 = -   - 15
(Sign is negative.) ――――
                    21 = (-21)

S1. -94 + 48 =

S2. 15 - -7 =

1. -42 + -63 =

2. -95 + 110 =

3. -9 - 12 =

4. 40 - -20 =

5. -12 + 5 + -16 =

6. 20 + -7 + 4 + -8 =

7. 64 - 93 =

8. 7 - -12 =

9. -425 + 501 =

10. -723 - 201 =

1.	
2.	
3.	
4.	
5.	
6.	
7.	
8.	
9.	
10.	
Score	

Problem Solving

If the temperature was -12° at midnight and by 6:00 A.M. it had dropped another 22°, what was the temperature at 6:00 A.M.?

Review Exercises

1. 336 + -521 =

2. -75 - 96 =

3. -402 + 763 =

4. 91 - -65 =

5. -256 + -758 =

6. -7 - 9 - 6 =

| **Helpful Hints** | Use what you have learned to solve the following problems. If you need help, refer to the examples on the previous page. |

S1. -763 - 202 =

S2. 95 - -62 =

1. -29 - 36 =

2. -428 + 500 =

3. 50 - -21 =

4. 72 - 125 =

5. 65 + 12 + -52 + 16 =

6. -37 + -16 + -42 =

7. -95 - 24 =

8. -55 - -30 =

9. -33 - 35 =

10. -316 + -422 =

1.
2.
3.
4.
5.
6.
7.
8.
9.
10.
Score

Problem Solving If the temperature was -12° at 6:00 A.M. and by noon it had risen 33°, what was the temperature at noon?

Review Exercises

1. 337
    98
    324
    +   7

2. 1,712
    -   963

3. 304
    x   27

4. 63⟌1724

5. -27 - 37 =

6. 27 - -37 =

Helpful Hints	The product of two integers with different signs is negative. The product of two integers with the same sign is positive. (• means multiply).	**Examples:** $7 • -16 = -$ (Sign is negative.) $\begin{array}{r} 16 \\ \times\ 7 \\ \hline 112 \end{array} = \boxed{-112}$ $-8 • -7 = +$ (Sign is positive.) $\begin{array}{r} 8 \\ \times\ 7 \\ \hline 56 \end{array} = \boxed{+56}$

S1. -4 x -12 =

S2. -17 • 8 =

1. -6 • -29 =

2. 16 • -5 =

3. -36 • -14 =

4. 27 • -22 =

5. -40 • 36 =

6. 9 x -19 =

7. -8 • -7 =

8. -24 • -16 =

9. 34 x -8 =

10. -17 • -38 =

1.

2.

3.

4.

5.

6.

7.

8.

9.

10.

Score

Problem Solving	Bill, Robert, and Olga together earned 520 dollars on Monday and 470 dollars on Tuesday. If they wanted to divide the money equally, how much would each person get?

Review Exercises

1. -6 • -7 =

2. 12 x -13 =

3. -15 • -20 =

4. 63 - 75 =

5. -66 + 95 =

6. -75 - -90 =

Helpful Hints	Use what you have learned to solve the following problems.
	Remember: The product of two integers with different signs is negative.
	The product of two integers with the same signs is positive.

S1. -15 • 12 =

S2. -16 • -20 =

1. -5 • -22 =

2. 5 x -22 =

3. -36 x -12 =

4. 42 x -10 =

5. -320 x 5 =

6. -75 • -5 =

7. 6 • -220 =

8. -18 • -40 =

9. 32 • -18 =

10. -160 • -15 =

1.

2.

3.

4.

5.

6.

7.

8.

9.

10.

Score

Problem Solving

A rope is 525 feet long. If it is cut into 5 pieces of equal length, how long will each piece be?

Review Exercises

1. -9 + 6 + -12 + 8 =

2. -7 - -9 =

3. -6 • -7 =

4. -6 x -42 =

5. -7 + -9 + -7 =

6. 16 - -20 =

Helpful Hints

When multiplying more than two integers, group them in pairs to simplify.

An integer next to parentheses means to multiply.

Examples:

2 • -3 (-6) =
(2 • -3) (-6) =
-6 (-6) = +
(Sign is positive.)

$$\begin{array}{r} 6 \\ \times\ 6 \\ \hline 36 \end{array} = \boxed{+36}$$

-2 • -3 • 4 • -2 =
(-2 • -3) • (4 • -2) =
6 • -8 = -
(Sign is negative.)

$$\begin{array}{r} 8 \\ \times\ 6 \\ \hline 48 \end{array} = \boxed{-48}$$

S1. -2 • 8 • -4 =

S2. (-2)(-5) • -3 =

1. 4 (-3) • 5 =

2. -3 • -6 (-7) =

3. 3 • -4 • -5 • 6 =

4. 5 (3) • -2 x (-6) =

5. 2 • -3 • -1 • -2 =

6. (-6) (-4) (-5) =

7. -3 (-4) • 1 (-4) =

8. 4 (-6) • 2 (-4) =

9. (-6) (-7) (2) (3) =

10. 11 (-10) (-4) =

1.	
2.	
3.	
4.	
5.	
6.	
7.	
8.	
9.	
10.	

Score

Problem Solving

If a plane can travel 460 miles per hour, how far can it travel in 8 hours?

Review Exercises

1. -72 + 96 =

2. -53 - 52 =

3. 56 + -62 - -50 =

4. 7 • -12 =

5. -2 (-6) (-3) =

6. (-2) (-3) (-4) =

Helpful Hints	Use what you have learned to solve the following problems. Remember, when multiplying more than two integers, group them into pairs to simplify.

S1. -9 • 6 • -7 =

S2. -10 (-6) • (-3) =

1. 4 (-5) (-6) =

2. -7 • -3 (-5) =

3. -2 • -3 • 4 • -6 =

4. 5 (6) • -5 • (-3) =

5. (-5) (-8) (-5) =

6. (2) (-3) (4) (-5) =

7. (-7) (-4) • 2 (-6) =

8. 5 (-6) • 4 (-3) =

9. (-6) (7) (-3) (2) =

10. -12 (3) (-6) =

1.

2.

3.

4.

5.

6.

7.

8.

9.

10.

Score

Problem Solving	A farmer owns 750 cows. If he decides to sell one-half of them, how many cows will he sell?

Review Exercises

1. -7 + 12 + -6 =

2. -7 - -9 =

3. 3 + -7 - -9 =

4. -63 - 72 - -60 =

5. 3 (-3) (-2) =

6. -5 (2) (2) (-3) =

| **Helpful Hints** | The quotient of two integers with different signs is negative.

The quotient of two integers with the same sign is positive. (Hint: determine the sign, then divide.). | **Examples:**

$36 \div -4 = -$ (Sign is negative.) $\quad \begin{array}{r} 9 \\ 4\overline{)36} \\ \underline{-36} \\ 0 \end{array} = \boxed{-9}$ $\quad \dfrac{-123}{-3} = +$ (Sign is positive.) $\quad \begin{array}{r} 41 \\ 3\overline{)123} \\ \underline{-12} \\ 3 \end{array} = \boxed{+4}$ |

S1. 12 ÷ -4 =

S2. $\dfrac{-75}{-5}$

1. -72 ÷ 4 =

2. 336 ÷ -7 =

3. $\dfrac{-105}{-5}$

4. 204 ÷ -4 =

5. $\dfrac{-130}{5}$

6. 576 ÷ -12 =

7. 56 ÷ -7 =

8. 357 ÷ -21 =

9. $\dfrac{-65}{-5}$

10. -684 ÷ -36 =

1.	
2.	
3.	
4.	
5.	
6.	
7.	
8.	
9.	
10.	
Score	

| **Problem Solving** | Eva is inviting 70 people to a party. She plans to provide one soft drink for each person invited. If soft drinks come in packs of six, how many packs must she buy? |

Review Exercises

1. $7\overline{)23}$ 2. $7\overline{)1396}$ 3. $60\overline{)896}$

4. $60\overline{)394}$ 5. $32\overline{)186}$ 6. $32\overline{)489}$

Helpful Hints	Use what you have learned to solve the following problems. The quotient of two integers with different signs is negative. The quotient of two integers with the same sign is positive.

S1. $96 \div -3 =$ S2. $\dfrac{-90}{-15}$ 1. $-72 \div 4 =$

2. $324 \div -4 =$ 3. $\dfrac{-115}{5}$ 4. $\dfrac{-2121}{-3}$

5. $864 \div -12 =$ 6. $\dfrac{-110}{-5}$ 7. $-80 \div -5 =$

8. $\dfrac{65}{-13}$ 9. $-104 \div 4 =$ 10. $-2001 \div -3 =$

1.
2.
3.
4.
5.
6.
7.
8.
9.
10.

Score

Problem Solving	A car traveled 295 miles in 5 hours. What was the car's average speed?

Review Exercises

1. 32 (-3) =

2. (-2) (-6) =

3. (-3) 2 (-4) =

4. 200 ÷ -4 =

5. $\dfrac{60}{-15}$

6. -90 ÷ -15 =

Helpful Hints	Use what you have learned to solve problems like these.	**Examples:** $\dfrac{-36 \div -9}{4 \div -2} = \dfrac{4}{-2} = \boxed{-2}$ (Sign is negative.) $\qquad \dfrac{4 \times -8}{-8 \div 2} = \dfrac{-32}{-4} = \boxed{+8}$ (Sign is positive.)

S1. $\dfrac{-15 \div -3}{5 \div -1} =$

S2. $\dfrac{3 \cdot (-8)}{-8 \div -4} =$

1. $\dfrac{-32 \div 4}{2 \cdot -2} =$

2. $\dfrac{-6 \times -5}{-30 \div -3} =$

3. $\dfrac{4 \cdot (-6)}{(-2) (-3)} =$

4. $\dfrac{-4 \cdot -9}{8 \div -4} =$

5. $\dfrac{-36 \div -6}{-10 \div 5} =$

6. $\dfrac{-24 \div -3}{-2 \cdot -2} =$

7. $\dfrac{75 \div -25}{3 \div -1} =$

8. $\dfrac{-42 \div -2}{14 \div -2} =$

9. $\dfrac{45 \div 5}{-9 \div 3} =$

10. $\dfrac{-56 \div -7}{-36 \div -9} =$

1.

2.

3.

4.

5.

6.

7.

8.

9.

10.

Score

Problem Solving	Kale has a library book that is seven days overdue. For each day overdue he must pay 25 cents. How much must he pay the library for his overdue book?

Review Exercises

1. Find the difference between 709 and 688.

2. Find the product of 42 and 604.

3. Find the quotient of 612 and 3.

4. $-77 - -60 =$

5. $(-3)(-2)(2) =$

6. $\dfrac{36 \div -9}{-8 \div -4} =$

Helpful Hints	Use what you have learned to solve problems. If necessary, refer to the examples on the previous page.

S1. $\dfrac{32 \div -4}{-8 \div 2} =$

S2. $\dfrac{5 \cdot -6}{-12 \div -4} =$

1. $\dfrac{2 \cdot -4}{-12 \div -6} =$

2. $\dfrac{-8 \text{ x } -5}{15 \div -3} =$

3. $\dfrac{4 \cdot -9}{-18 \div -3} =$

4. $\dfrac{-4 \cdot -20}{-32 \div 4} =$

5. $\dfrac{-54 \div -9}{-2 \cdot -3} =$

6. $\dfrac{40 \div -5}{-20 \div 5} =$

7. $\dfrac{10 \cdot -3}{-3 \cdot -5} =$

8. $\dfrac{32 \div -2}{-2 \cdot -2} =$

9. $\dfrac{-55 \div 11}{-25 \div -5} =$

10. $\dfrac{-42 \div 7}{-12 \div -4} =$

1.
2.
3.
4.
5.
6.
7.
8.
9.
10.

Score

Problem Solving	Elena scored a total of 475 points on five tests. What was her average score?

Review Exercises

1. 77 + 99 + 73 + 62 =

2. 7,165
    769
    + 963

3.   308
    x 206

4.   72,172
    - 15,564

5. 6 x 1,096 =

6. 6,102 - 1,765 =

Helpful Hints

Use what you have learned to solve the following problems.
Remember to be careful with positive and negative signs.

| | 1. |
| | 2. |

S1. (-5) • 3 (-4) x 6 =

S2. $\dfrac{20 \div -2}{-25 \div 5}$ =

1. 15 • -6 =

| | 3. |
| | 4. |

2. 6 x -5 • 4 =

3. (-3) (2) (-4) =

4. $\dfrac{-75}{-5}$

| | 5. |
| | 6. |

5. 32 ÷ -4 =

6. -224 ÷ -4 =

7. (-3) (-2) (6) =

	7.
	8.
	9.

8. $\dfrac{-30 \div -5}{2 \cdot -3}$ =

9. $\dfrac{20 \times -3}{-50 \div -10}$ =

10. $\dfrac{36 \div -9}{-16 \div 4}$ =

| | 10. |

Score

Problem Solving

Fariba had test scores of 80, 84, 96, and 100.
What was her average score?

Review Exercises

1. $3\overline{)76}$

2. $\dfrac{1250}{5}$

3. $41\overline{)361}$

4. $41\overline{)765}$

5. $28\overline{)349}$

6. $28\overline{)2106}$

Helpful Hints	Use what you have learned to solve the following problems. Sometimes it is helpful to review the examples from previous pages if you need help.

S1. $-3 \cdot 4 \cdot -4 \cdot 3 =$

S2. $\dfrac{15 \cdot -2}{-2 \cdot -3}$

1. $25 \cdot (-6) =$

2. $5 \cdot (-6)(-3) =$

3. $(5)(-4)(2)(-2) =$

4. $\dfrac{-125}{-5}$

5. $54 \div -6 =$

6. $-111 \div -3 =$

7. $\dfrac{(-3)(-6)}{-9}$

8. $\dfrac{60 \div -6}{-2 \cdot -5} =$

9. $\dfrac{-72 \div 9}{-20 \div -5} =$

10. $\dfrac{20 \times -3}{-4 \times -5} =$

1. _____

2. _____

3. _____

4. _____

5. _____

6. _____

7. _____

8. _____

9. _____

10. _____

Score

Problem Solving	The attendance at the Eagle's game last year was 42,728. This year 53,962 attended. What was the increase in attendance this year?

Review of All Integer Operations

1. -8 + 5 =

2. 8 + -5 =

3. -8 + -5 =

4. -6 + -8 + 17 =

5. -35 + 19 + 23 + -33 =

6. 6 - 8 =

7. 4 - -8 =

8. -3 - 7 =

9. -12 - 16 =

10. 18 - 19 =

11. 4 • -15 =

12. -3 • -27 =

13. 3 (-6) (-4) =

14. (-2) • 5 (-6) • 2 =

15. -27 ÷ 9 =

16. -234 ÷ -3 =

17. $\dfrac{-256}{-8}$

18. $\dfrac{-24 \div 2}{18 \div 3}$

19. $\dfrac{8 \cdot (-4)}{-20 \div -5}$

20. $\dfrac{-30 \cdot -2}{-60 \div -10}$

1.	
2.	
3.	
4.	
5.	
6.	
7.	
8.	
9.	
10.	
11.	
12.	
13.	
14.	
15.	
16.	
17.	
18.	
19.	
20.	

Final Review of All Integer Operations

1. 12 + -6 =

2. -12 + 6 =

3. -12 + -6 =

4. -6 + -9 + 21 =

5. 65 + -32 + 15 + -50 =

6. 9 - 12 =

7. 15 - -16 =

8. -20 - 31 =

9. -23 - 26 =

10. 16 - 80 =

11. -16 • 7 =

12. -15 • -10 =

13. 3 (-12) (-4) =

14. (-2) 3 (-5) 3 =

15. $\dfrac{-72}{-8}$

16. -414 ÷ -3 =

17. $\dfrac{884}{-4}$

18. $\dfrac{80 ÷ -8}{-25 ÷ 5}$

19. $\dfrac{-12 • -3}{-2 • 2}$

20. $\dfrac{75 ÷ -25}{-3 ÷ -1}$

1.	
2.	
3.	
4.	
5.	
6.	
7.	
8.	
9.	
10.	
11.	
12.	
13.	
14.	
15.	
16.	
17.	
18.	
19.	
20.	

Whole Numbers - Final Test

1. $\begin{array}{r} 356 \\ + 397 \\ \hline \end{array}$

2. $\begin{array}{r} 623 \\ 462 \\ 17 \\ + 816 \\ \hline \end{array}$

3. $6,502 + 916 + 15,989 =$

4. $6,667 + 3,444 + 755 + 16 =$

5. $7,093 + 675 + 97 + 768 =$

6. $\begin{array}{r} 916 \\ - 428 \\ \hline \end{array}$

7. $\begin{array}{r} 5,392 \\ - 1,768 \\ \hline \end{array}$

8. $7,053 - 4,289 =$

9. $5,000 - 3,296 =$

10. $7,008 - 999 =$

11. $\begin{array}{r} 87 \\ \times \ 4 \\ \hline \end{array}$

12. $\begin{array}{r} 7,132 \\ \times \quad 4 \\ \hline \end{array}$

13. $\begin{array}{r} 45 \\ \times \ 36 \\ \hline \end{array}$

14. $\begin{array}{r} 392 \\ \times \ 47 \\ \hline \end{array}$

15. $\begin{array}{r} 743 \\ \times \ 247 \\ \hline \end{array}$

16. $4\overline{)626}$

17. $4\overline{)1428}$

18. $40\overline{)568}$

19. $30\overline{)8672}$

20. $18\overline{)1343}$

1.
2.
3.
4.
5.
6.
7.
8.
9.
10.
11.
12.
13.
14.
15.
16.
17.
18.
19.
20.

Integers - Final Test

1. $7 + -5 =$ 2. $-7 + 5 =$ 3. $-7 + -5 =$

4. $-7 + 4 + -3 + 6 =$ 5. $-52 + 17 + 22 + -21 =$

6. $7 - 9 =$ 7. $4 - -7 =$ 8. $-7 - 12 =$

9. $-12 - 16 =$ 10. $14 - 19 =$ 11. $4 \cdot (-8) =$

12. $-12 \cdot -19 =$ 13. $4 (-5) (-6) =$ 14. $-2 \cdot 3 \cdot (-7) =$

15. $-63 \div 9 =$ 16. $-560 \div -5 =$ 17. $\dfrac{-136}{-8}$

18. $\dfrac{32 \div -2}{-16 \div -4}$ 19. $\dfrac{-8 \cdot -5}{-15 \div 3}$ 20. $\dfrac{-40 \cdot -2}{-20 \div 2}$

1.	
2.	
3.	
4.	
5.	
6.	
7.	
8.	
9.	
10.	
11.	
12.	
13.	
14.	
15.	
16.	
17.	
18.	
19.	
20.	

Answer Key

PAGE 8
Review Exercises:
1. 357
2. 627
3. 13
4. 739
5. 16
6. 49
S1. 819
S2. 933
1. 113
2. 783
3. 1,000
4. 1,072
5. 1,497
6. 687
7. 1,168
8. 120
9. 1,673
10. 412
Problem Solving: 134 students

PAGE 9
Review Exercises:
1. 449
2. 731
3. 83
4. 1,093
5. 1,178
6. 944
S1. 1,285
S2. 734
1. 1,264
2. 135
3. 1,767
4. 675
5. 1,913
6. 1,551
7. 329
8. 1,180
9. 288
10. 1,578
Problem Solving: $220

PAGE 10
Review Exercises:
1. 191
2. 1,225
3. 2,080
4. 1,906
5. 137
6. 62
S1. 7,619
S2. 44,027
1. 10,395
2. 24,708
3. 3,628,991
4. 7,497,988
5. 71,078
6. 94,180
7. 7,697
8. 105,598
9. 118,112
10. 899,935
Problem Solving: 36,283,995

PAGE 11
Review Exercises:
1. 885
2. 53,203
3. 134,680
4. 33,104,468
5. 211,628
6. 10,562,592
S1. 100,463
S2. 6,301,851
1. 78,734
2. 8,976,984
3. 12,915
4. 86,835
5. 20,422
6. 1,166,094
7. 212,299
8. 14,362,605
9. 97,625
10. 42,865
Problem Solving: 8,044 students

PAGE 12
Review Exercises:
1. 154
2. 43,939
3. 19,731,174
4. 1,615
5. 1,370,727
6. 337
S1. 384
S2. 2,181
1. 315
2. 571
3. 5,395
4. 2,775
5. 648
6. 5,098
7. 716
8. 3,389
9. 6,470
10. 491
Problem Solving: 97 students

PAGE 13
Review Exercises:
1. 1,883
2. 278
3. 6,057
4. 2,126
5. 4,861
6. 41,989,515
S1. 559
S2. 7,366
1. 473
2. 175
3. 1,885
4. 4,159
5. 4,667
6. 6,279
7. 11,874
8. 1,517
9. 2,369
10. 2,928
Problem Solving: $58

Whole Numbers and Integers—Solutions

PAGE 14
Review Exercises:
1. 1,199
2. 6,766
3. 3,451
4. 8,682
5. 1,263
6. 2,656
S1. 245
S2. 433
1. 23
2. 529
3. 2,667
4. 462
5. 7,242
6. 3,668
7. 2,216
8. 1,337
9. 6,678
10. 7,485
Problem Solving: 291 seats

PAGE 15
Review Exercises:
1. 224
2. 4,214
3. 277
4. 4,249
5. 3,979
6. 4,232
S1. 4,546
S2. 6,033
1. 823
2. 2,428
3. 664
4. 4,803
5. 1,649
6. 42,458
7. 68,296
8. 4,494
9. 443
10. 47,192
Problem Solving: $16,500

PAGE 16
Review Exercises:
1. 830
2. 51,656
3. 1,969
4. 4,691
5. 4,236
6. 1,827
S1. 24,907
S2. 2,674
1. 632
2. 461
3. 13,049
4. 2,211
5. 4,303
6. 165
7. 9,979
8. 992
9. 778
10. 152,371
Problem Solving: $95

PAGE 17
Review Exercises:
1. 96
2. 73,212
3. 1,251
4. 64,738
5. 2,474
6. 79,922
S1. 5,655
S2. 52,811
1. 676
2. 5,764
3. 10,808
4. 27,891
5. 198
6. 344
7. 1,793
8. 67,384
9. 4,509
10. 37,046
Problem Solving: 8603 people

PAGE 18
Review Exercises:
1. 234
2. 1,052
3. 2,335
4. 148
5. 7,216
6. 40,117
S1. 1,578
S2. 19,524
1. 201
2. 444
3. 2,562
4. 25,284
5. 24,448
6. 43,536
7. 49,252
8. 15,396
9. 52,572
10. 34,848
Problem Solving: 2,555 days

PAGE 19
Review Exercises:
1. 1,372
2. 21,448
3. 2,803
4. 4,765
5. 874
6. 3,052
S1. 43,404
S2. 72,056
1. 4,368
2. 81,108
3. 166,082
4. 45,792
5. 87,570
6. 28,036
7. 21,144
8. 56,084
9. 142,368
10. 74,574
Problem Solving: 152 points

Whole Numbers and Integers—Solutions

PAGE 20
Review Exercises:
1. 4,350
2. 27,108
3. 2,010
4. 15,846
5. 3,327
6. 1,462
S1. 1,032
S2. 13,038
1. 4,346
2. 782
3. 3,995
4. 10,900
5. 16,800
6. 33,182
7. 592
8. 4,324
9. 34,335
10. 15,180
Problem Solving: 875 desks

PAGE 21
Review Exercises:
1. 2,839
2. 22,015
3. 1,728
4. 10,575
5. 15,726
6. 5,563
S1. 10,452
S2. 15,664
1. 1,300
2. 1,288
3. 2,530
4. 13,824
5. 18,848
6. 35,700
7. 9,476
8. 12,168
9. 4,550
10. 44,384
Problem Solving: 1200 girls

PAGE 22
Review Exercises:
1. 1,512
2. 11,396
3. 14,000
4. 6,255
5. 154
6. 4,164
S1. 64,719
S2. 380,824
1. 77,805
2. 33,702
3. 73,688
4. 183,092
5. 224,802
6. 107,600
7. 175,577
8. 67,808
9. 411,000
10. 162,504
Problem Solving: 158,600 cars

PAGE 23
Review Exercises:
1. 11,853
2. 824
3. 13,032
4. 1,058
5. 7,890
6. 370,304
S1. 99,900
S2. 317,100
1. 649,602
2. 371,500
3. 140,505
4. 489,240
5. 662,592
6. 229,272
7. 88,914
8. 427,000
9. 692,804
10. 612,381
Problem Solving: 784 students

PAGE 24
Review Exercises:
1. 2,448
2. 864
3. 25,334
4. 777
5. 4,452
6. 11,016
S1. 10,650
S2. 260,525
1. 185
2. 4,230
3. 16,422
4. 4,176
5. 15,435
6. 70,016
7. 10,890
8. 459,900
9. 20,130
10. 125,356
Problem Solving: 10,080 minutes

PAGE 25
Review Exercises:
1. 726
2. 3,135
3. 6,329
4. 1,498
5. 15,862
6. 182,000
S1. 30,035
S2. 19,085
1. 2,020
2. 42,012
3. 62,412
4. 7,680
5. 45,234
6. 307,824
7. 2,160
8. 70,704
9. 47,905
10. 158,704
Problem Solving: 101 people

Whole Numbers and Integers—Solutions

PAGE 26

Review Exercises:
1. 307
2. 1,735
3. 88,625
4. 6,917
5. 2,992
6. 61,800
S1. 4 r1
S2. 9 r4
1. 24 r2
2. 5 r7
3. 13
4. 18 r3
5. 12
6. 15 r4
7. 9 r5
8. 24
9. 16 r1
10. 23 r1
Problem Solving: 18 pencils

PAGE 27

Review Exercises:
1. 8 r1
2. 17 r4
3. 6 r3
4. 106
5. 1155
6. 1551
S1. 18 r1
S2. 4 r3
1. 9 r2
2. 15
3. 11 r5
4. 48 r1
5. 8 r4
6. 5 r4
7. 15 r2
8. 24 r3
9. 8 r1
10. 13
Problem Solving: $78,000

PAGE 28

Review Exercises:
1. 6 r2
2. 13 r4
3. 4,486
4. 3,682
5. 3,772
6. 222
S1. 261 r1
S2. 45 r4
1. 324
2. 209 r1
3. 133 r2
4. 121 r1
5. 114 r4
6. 141 r4
7. 312
8. 121 r2
9. 258 r2
10. 89 r5
Problem Solving: 36 packages

PAGE 29

Review Exercises:
1. 19
2. 12 r5
3. 123 r1
4. 74 r4
5. 132
6. 64 r2
S1. 187 r1
S2. 94 r1
1. 114 r4
2. 78 r6
3. 97
4. 122
5. 127
6. 41 r2
7. 133 r1
8. 99
9. 155 r3
10. 101 r3
Problem Solving: 5,400 miles

PAGE 30

Review Exercises:
1. 23
2. 133 r2
3. 10,011
4. 5,373,870
5. 552,600
6. 773
S1. 3017 r1
S2. 670 r2
1. 4,617 r1
2. 3,681
3. 1,136 r5
4. 448
5. 1,533 r1
6. 1,780 r3
7. 12,247
8. 12,181
9. 10,309
10. 4,402 r4
Problem Solving: $3,310

PAGE 31

Review Exercises:
1. 175 r1
2. 201 r2
3. 376 r2
4. 1,235
5. 2,208 r4
6. 6,332
S1. 662 r4
S2. 2,795 r1
1. 1,374 r1
2. 345 r1
3. 1,112 r6
4. 887 r2
5. 1551 r3
6. 1,002
7. 5,574 r1
8. 3,186 r3
9. 15,263 r1
10. 4,714 r4
Problem Solving: 44 miles

PAGE 32

Review Exercises:
1. 782
2. 25,368
3. 11,102
4. 125
5. 193 r5
6. 119,507
S1. 300 r1
S2. 1,080 r3
1. 80 r2
2. 1,700
3. 2,103 r1
4. 445 r6
5. 254 r2
6. 1,200
7. 2,002 r1
8. 973 r8
9. 904 r1
10. 804 r1
Problem Solving: 12,500 sheets

PAGE 33

Review Exercises:
1. 2,459
2. 319,774
3. 85 r5
4. 1,000 r3
5. 1,525 r1
6. 281 r2
S1. 123
S2. 1,462
1. 61 r2
2. 17 r4
3. 202 r1
4. 1,001 r5
5. 1,622 r4
6. 349 r2
7. 1,677 r7
8. 2,877 r2
9. 11,872 r2
10. 3,071
Problem Solving: 556 containers

PAGE 34

Review Exercises:
1. 119 r4
2. 1,142
3. 226,800
4. 1,001 r2
5. 233 r1
6. 16,253
S1. 5 r26
S2. 143 r12
1. 3 r28
2. 7 r26
3. 8 r38
4. 11 r59
5. 56 r2
6. 69 r16
7. 224 r12
8. 101 r6
9. 94 r26
10. 59 r36
Problem Solving: 256 boxes

PAGE 35

Review Exercises:
1. 34 r1
2. 1,002 r5
3. 1,577
4. 10,944
5. 5,132
6. 2,041
S1. 2 r37
S2. 126 r52
1. 19 r16
2. 9 r7
3. 77 r2
4. 246 r16
5. 81 r54
6. 142 r48
7. 199 r26
8. 44 r6
9. 27 r49
10. 241 r26
Problem Solving: 768 students

PAGE 36

Review Exercises:
1. 1,046
2. 5,768
3. 250
4. 12 r9
5. 5 r17
6. 131 r36
S1. 21 r1
S2. 8 r23
1. 22 r34
2. 22 r1
3. 12 r1
4. 9 r15
5. 24 r10
6. 31 r15
7. 12 r15
8. 11 r26
9. 44 r12
10. 20 r26
Problem Solving: 14 gallons

PAGE 37

Review Exercises:
1. 11,410
2. 162,000
3. 206
4. 2,244
5. 16 r26
6. 81 r82
S1. 22 r10
S2. 7 r10
1. 13 r30
2. 22 r18
3. 13 r2
4. 36 r4
5. 32 r9
6. 24 r18
7. 10 r46
8. 31 r2
9. 34 r24
10. 17 r36
Problem Solving: 185 seats

Whole Numbers and Integers—Solutions

PAGE 38
Review Exercises:
1. 3 r8
2. 13 r13
3. 8 r3
4. 13 r10
5. 8 r19
6. 14 r13
S1. 3 r71
S2. 19 r4
1. 5 r7
2. 5 r82
3. 54 r15
4. 8 r2
5. 5 r3
6. 63 r4
7. 19 r3
8. 19 r31
9. 20 r1
10. 43 r4
Problem Solving: 900 parts

PAGE 39
Review Exercises:
1. 78
2. 17,332
3. 1,472
4. 5,175
5. 250,800
6. 388,367
S1. 29 r21
S2. 37 r16
1. 31 r14
2. 32 r5
3. 30 r2
4. 20 r16
5. 29 r28
6. 49
7. 43 r13
8. 19 r31
9. 63 r4
10. 8 r10
Problem Solving: $53.00

PAGE 40
Review Exercises:
1. 365
2. 467
3. 30,658
4. 1,765
5. 2,969
6. 1,347
S1. 215 r5
S2. 207 r11
1. 125 r20
2. 148 r15
3. 214 r17
4. 165 r11
5. 298 r10
6. 306 r1
7. 32 r62
8. 80 r5
9. 142
10. 750 r6
Problem Solving: 440 miles

PAGE 41
Review Exercises:
1. 35 r1
2. 872 r1
3. 107 r42
4. 42 r27
5. 12 r2
6. 8 r33
S1. 48 r5
S2. 212 r11
1. 94 r22
2. 77 r35
3. 114 r14
4. 94 r28
5. 116 r5
6. 51 r54
7. 317 r17
8. 163 r19
9. 401 r1
10. 43 r3
Problem Solving:
11 cartons, 27 left

PAGE 42
Review Exercises:
1. 107,083
2. 4,329
3. 155,935
4. 218,225
5. 5,110
6. 7,392
S1. 95 r21
S2. 55 r 21
1. 38 r1
2. 331 r3
3. 279 r2
4. 16 r7
5. 7 r76
6. 41 r72
7. 25 r16
8. 12 r2
9. 54 r50
10. 253
Problem Solving: 51 students

PAGE 43
Review Exercises:
1. 34
2. 654 r2
3. 9 r9
4. 45 r17
5. 11 r25
6. 73 r11
S1. 96 r16
S2. 77 r17
1. 19 r2
2. 325 r5
3. 778
4. 16 r7
5. 6 r36
6. 84 r56
7. 5 r16
8. 12 r50
9. 23 r48
10. 130 r28
Problem Solving: $960

PAGE 44
1. 1,109
2. 2,899
3. 8,1387
4. 10,675
5. 8,474
6. 345
7. 2,815
8. 6,285
9. 4,012
10. 6,285
11. 261
12. 45,794
13. 3,087
14. 61,484
15. 113,702
16. 141 r2
17. 282 r5
18. 21 r27
19. 150 r26
20. 44 r26

PAGE 45
1. 1,846
2. 89,844
3. 32,904
4. 8,638
5. 26,297
6. 435
7. 8,215
8. 44,499
9. 2,643
10. 55,807
11. 2,156
12. 63,090
13. 2,523
14. 38,142
15. 771,363
16. 58 r6
17. 232
18. 5 r39
19. 142 r28
20. 594 r10

PAGE 46
Review Exercises:
1. 1,840
2. 6,894
3. 3,783
4. 233 r6
5. 5 r17
6. 17 r39
S1. 6
S2. -23
1. 9
2. -18
3. -16
4. -27
5. 7
6. 38
7. -180
8. -29
9. -70
10. -20
Problem Solving: 6 buses

PAGE 47
Review Exercises:
1. -14
2. 5
3. -78
4. 148,131
5. 12 r25
6. 6 r5
S1. 9
S2. -185
1. -9
2. -936
3. -37
4. -141
5. 37
6. -215
7. -901
8. -99
9. -641
10. 154
Problem Solving: $31

PAGE 48
Review Exercises:
1. 21,063
2. 1,257
3. 58,398
4. -15
5. -188
6. 37
S1. -2
S2. -5
1. 3
2. -3
3. -6
4. 13
5. -6
6. -8
7. -33
8. -9
9. -26
10. -89
Problem Solving: 89

PAGE 49
Review Exercises:
1. 24 r7
2. 9 r24
3. 21 r25
4. 62 r3
5. 2
6. -72
S1. -6
S2. 9
1. -7
2. -22
3. -55
4. 14
5. -34
6. 14
7. -24
8. -108
9. 33
10. -57
Problem Solving: 24 packages

Whole Numbers and Integers—Solutions

PAGE 50

Review Exercises:
1. -28
2. 8
3. -15
4. 241
5. 1,229
6. 17,292
S1. -13
S2. 1
1. 15
2. -3
3. 8
4. -31
5. 66
6. 34
7. -20
8. -18
9. -93
10. 31
Problem Solving: -8°

PAGE 51

Review Exercises:
1. -9
2. -10
3. -30
4. -42
5. 20
6. -16
S1. -28
S2. 112
1. 38
2. -8
3. 12
4. -46
5. 71
6. -79
7. -14
8. 48
9. -434
10. -16
Problem Solving: $195

PAGE 52

Review Exercises:
1. -16
2. -2
3. 40
4. 21
5. -4
6. -107
S1. -46
S2. 22
1. -105
2. 15
3. -21
4. 60
5. -23
6. 9
7. -29
8. 19
9. 76
10. -924
Problem Solving: -34

PAGE 53

Review Exercises:
1. -185
2. -171
3. 361
4. 156
5. -1,014
6. -22
S1. -965
S2. 157
1. -65
2. 72
3. 71
4. -53
5. 41
6. -95
7. -119
8. -25
9. -68
10. -738
Problem Solving: 21°

PAGE 54

Review Exercises:
1. 766
2. 749
3. 8,208
4. 27 r23
5. -64
6. 64
S1. 48
S2. -136
1. 174
2. -80
3. 504
4. -594
5. -1440
6. -171
7. 56
8. 384
9. -272
10. 646
Problem Solving: $330

PAGE 55

Review Exercises:
1. 42
2. -156
3. 300
4. -12
5. 29
6. 15
S1. -180
S2. 320
1. 110
2. -110
3. 432
4. -420
5. -1,600
6. 375
7. -1,320
8. 720
9. -576
10. 2,400
Problem Solving: 105 feet

Whole Numbers and Integers—Solutions

PAGE 56
Review Exercises:
1. -7
2. 2
3. 42
4. 252
5. -23
6. 36
S1. 64
S2. -30
1. -60
2. -126
3. 360
4. 180
5. -12
6. -120
7. -48
8. 192
9. 252
10. 440
Problem Solving: 3,680 miles

PAGE 57
Review Exercises:
1. 24
2. -105
3. 44
4. -84
5. -36
6. -24
S1. 378
S2. -180
1. 120
2. -105
3. -144
4. 450
5. -200
6. 120
7. -336
8. 360
9. 252
10. 216
Problem Solving: 375 cows

PAGE 58
Review Exercises:
1. -1
2. 2
3. 5
4. -75
5. 18
6. 60
S1. -3
S2. 15
1. -18
2. -48
3. 21
4. -51
5. -26
6. -48
7. -8
8. -17
9. 13
10. 19
Problem Solving: 12 packs

PAGE 59
Review Exercises:
1. 3 r2
2. 199 r3
3. 14 r56
4. 6 r34
5. 5 r26
6. 15 r9
S1. -32
S2. 6
1. -18
2. -81
3. -23
4. 707
5. -72
6. 22
7. 16
8. -5
9. -26
10. 667
Problem Solving:
    59 miles per hour

PAGE 60
Review Exercises:
1. -96
2. 12
3. 24
4. -50
5. -4
6. 6
S1. -1
S2. -12
1. 2
2. 3
3. -4
4. -18
5. -3
6. 2
7. 1
8. -3
9. -3
10. 2
Problem Solving: $1.75

PAGE 61
Review Exercises:
1. 21
2. 25,368
3. 204
4. -17
5. 12
6. -2
S1. 2
S2. -10
1. -4
2. -8
3. -6
4. -10
5. 1
6. 2
7. -2
8. -4
9. -1
10. -2
Problem Solving: 95

Whole Numbers and Integers—Solutions

PAGE 62
Review Exercises:
1. 311
2. 8,897
3. 63,448
4. 56,608
5. 6,576
6. 4,337
S1. 360
S2. 2
1. -90
2. -120
3. 24
4. 15
5. -8
6. 56
7. 36
8. -1
9. -12
10. 1
Problem Solving: 90

PAGE 63
Review Exercises:
1. 25 r1
2. 250
3. 8 r33
4. 18 r27
5. 12 r13
6. 75 r6
S1. 144
S2. -5
1. -150
2. 90
3. 80
4. 25
5. -9
6. 37
7. -2
8. -1
9. -2
10. -3
Problem Solving: 11,234

PAGE 64
1. -3
2. 3
3. -13
4. 3
5. -26
6. -2
7. 12
8. -10
9. -28
10. -1
11. -60
12. 81
13. 72
14. 120
15. -3
16. 78
17. 32
18. -2
19. -8
20. 10

PAGE 65
1. 6
2. -6
3. -18
4. 6
5. -2
6. -3
7. 31
8. -51
9. -49
10. -64
11. -112
12. 150
13. 144
14. 90
15. 9
16. 138
17. -221
18. 2
19. -9
20. -1

PAGE 66
1. 753
2. 1,918
3. 23,407
4. 10,882
5. 8,633
6. 488
7. 3,624
8. 2,764
9. 1,704
10. 6,009
11. 348
12. 28,528
13. 1,620
14. 18,424
15. 183,521
16. 156 r2
17. 357
18. 14 r8
19. 289 r2
20. 74 r11

PAGE 67
1. 2
2. -2
3. -12
4. 0
5. -34
6. -2
7. 11
8. -19
9. -28
10. -5
11. -32
12. 228
13. 120
14. 42
15. -7
16. 112
17. 17
18. -4
19. -8
20. -8

Math Resource Center

A

absolute value The distance of a number from 0 on the number line. The absolute value is always positive.

acute angle An angle with a measure of less than 90 degrees.

adjacent Next to.

algebraic expression A mathematical expression that contains at least one variable.

angle Any two rays that share an endpoint will form an angle.

associative properties For any a, b, c:
addition: $(a + b) + c = a + (b + c)$
multiplication: $(ab)c = a(bc)$

B

base The number being multiplied. In an expression such as 4^2, 4 is the base.

C

coefficient A number that multiplies the variable. In the term 7x, 7 is the coefficient of x.

commutative properties For any a, b:
addition: $a + b = b + a$
multiplication: $ab = ba$

complementary angles Two angles that have measures whose sum is 90 degrees.

congruent Two figures having exactly the same size and shape.

coordinate plane The plane which contains the x- and y-axes. It is divided into 4 quadrants. Also called coordinate system and coordinate grid.

coordinates An ordered pair of numbers that identify a point on a coordinate plane.

D

data Information that is organized for analysis.

degree A unit that is used in measuring angles.

denominator The bottom number of a fraction that tells the number of equal parts into which a whole is divided.

disjoint sets Sets that have no members in common. {1,2,3} and {4,5,6} are disjoint sets.

Glossary

distributive property   For real numbers a, b, and c: a(b + c) = ab + ac.

E

element of a set   Member of a set.

empty set   The set that has no members. Also called the null set and written Ø or { }.

equation   A mathematical sentence that contains an equal sign (=) and states that one expression is equal to another expression.

equivalent   Having the same value.

exponent   A number that indicates the number of times a given base is used as a factor. In the expression n^2, 2 is the exponent.

expression   Variables, numbers, and symbols that show a mathematical relationship.

extremes of a proportion   In the proportion $\frac{a}{b} = \frac{c}{d}$, a and d are the extremes.

F

factor   An integer that divides evenly into another.

finite   Something that is countable.

formula   A general mathematical statement or rule. Used often in algebra and geometry.

function   A set of ordered pairs that pairs each x-value with one and only one y-value.
(0,2), (-1,6), (4,-2), (-3,4) is a function.

G

graph   To show points named by numbers or ordered pairs on a number line or coordinate plane.
Also, a drawing to show the relationship between sets of data.

greatest common factor   The largest common factor of two or more numbers. Also written GCF.
The greatest common factor of 15 and 25 is 5.

grouping symbols   Symbols that indicate the order in which mathematical operations should take place.
Examples include parentheses ( ), brackets [ ], braces { }, and fraction bars —— .

H

hypotenuse   The side opposite the right angle in a right triangle.

I

identity properties of addition and multiplication   For any real number a:
addition: $a + 0 = 0 + a = a$
multiplication: $1 \times a = a \times 1 = a$

inequality   A mathematical sentence that states one expression is greater than or less than another.
Inequality symbols are read as follows: $<$ less than
$\leq$ less than or equal to
$>$ greater than
$\geq$ greater than or equal to

infinite   Having no boundaries or limits. Uncountable.

integers   Numbers in a set. ...-3, -2, -1, 0, 1, 2, 3...

intersection of sets   If A and B are sets, then A intersection B is the set whose members are included in both sets A and B, and is written $A \cap B$. If set A = {1,2,3,4} and set B = {1,3,5}, then $A \cap B$ = {1,3}

inverse properties of addition and multiplication   For any number a:
addition: $a + -a = 0$
multiplication: $a \times 1/a = 1$ $(a \neq 0)$

inverse operations   Operations that "undo" each other. Addition and subtraction are inverse operations, and multiplication and division are inverse operations.

L

least common multiple   The least common multiple of two or more whole numbers is the smallest whole number, other than zero, that they all divide into evenly. Also written as LCM. The least common multiple of 12 and 8 is 24.

linear equation   An equation whose graph is a straight line.

M

mean   In statistics, the sum of a set of numbers divided by the number of elements in the set.
Sometimes referred to as average.

means of a proportion   In the proportion $\frac{a}{b} = \frac{c}{d}$, b and c are the means.

median   In statistics, the middle number of a set of numbers when the numbers are arranged in order of least to greatest. If there are two middle numbers, find their mean.

mode   In statistics, the number that appears most frequently. Sometimes there is no mode. There may also be more than one mode.

multiple   The product of a whole number and another whole number.

Glossary

natural numbers   Numbers in the set 1, 2, 3, 4,... Also called counting numbers.

negative numbers   Numbers that are less than zero.

null set   The set that has no members. Also called the empty set and written Ø or { }.

number line   A line that represents numbers as points.

numerator   The top part of a fraction.

obtuse angle   An angle whose measure is greater than 90° and less than 180°.

opposites   Numbers that are the same distance from zero, but are on opposite sides of zero on a number line. 4 and -4 are opposites.

order of operations   The order of steps to be used when simplifying expressions.
1. Evaluate within grouping symbols.
2. Eliminate all exponents.
3. Multiply and divide in order from left to right.
4. Add and subtract in order from left to right.

ordered pair   A pair of numbers (x,y) that represent a point on the coordinate plane. The first number is the x-coordinate and the second number is the y-coordinate.

origin   The point where the x-axis and the y-axis intersect in a coordinate plane. Written as (0,0).

outcome   One of the possible events in a probability situation.

parallel lines   Lines in a plane that do not intersect. They stay the same distance apart.

percent   Hundredths or per hundred. Written %.

perimeter   The distance around a figure.

perpendicular lines   Lines in the same plane that intersect at a right (90°) angle.

pi   The ratio of the circumference of a circle to its diameter. Written π. The approximate value for π is 3.14 as a decimal and $\frac{22}{7}$ as a fraction.

plane   A flat surface that extends infinitely in all directions.

point   An exact position in space. Points also represent numbers on a number line or coordinate plane.

positive number   Any number that is greater than 0.

power   An exponent.

prime number   A whole number greater than 1 whose only factors are 1 and itself.

probability   What chance, or how likely it is for an event to occur. It is the ratio of the ways a certain outcome can occur and the number of possible outcomes.

proportion   An equation that states that two ratios are equal. $\frac{4}{8} = \frac{2}{4}$ is a proportion.

Pythagorean theorem   In a right triangle, if c is the hypotenuse, and a and b are the other two legs, then $a^2 + b^2 = c^2$.

Q

quadrant   One of the four regions into which the x-axis and y-axis divide a coordinate plane.

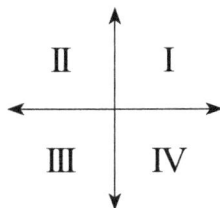

R

range   The difference between the greatest number and the least number in a set of numbers.

ratio   A comparison of two numbers using division. Written a:b, a to b, and a/b.

reciprocals   Two numbers whose product is 1. $\frac{2}{3}$ and $\frac{3}{2}$ are reciprocals because $\frac{2}{3} \times \frac{3}{2} = 1$.

reduce   To express a fraction in its lowest terms.

relation   Any set of ordered pairs.

right angle   An angle that has a measure of 90°.

rise   The change in y going from one point to another on a coordinate plane. The vertical change.

run   The change in x going from one point to another on a coordinate plane. The horizontal change.

S

scientific notation   A number written as the product of a numbers between 1 and 10 and a power of ten. In scientific notation, $7,000 = 7 \times 10^3$.

set   A well-defined collection of objects.

slope   Refers to the slant of a line. It is the ratio of rise to run.

Glossary

solution   A number that can be substituted for a variable to make an equation true.

square root   Written $\sqrt{}$. The $\sqrt{36} = 6$ because $6 \times 6 = 36$.

statistics   Involves data that is gathered about people or things and is used for analysis.

subset   If all the members of set A are members of set B, then set A is a subset of set B. Written $A \subset B$. If set A = {1,2,3} and set B = {0,1,2,3,5,8}, set A is a subset of set B because all of the members of a set A are also members of set B.

U

union of sets   If A and B are sets, the union of set A and set B is the set whose members are included in set A, or set B, or both set A and set B. A union B is written $A \cup B$. If set = {1,2,3,4} and set B = {1,3,5,7}, then $A \cup B$ = {1,2,3,4,5,7}.

universal set   The set which contains all the other sets which are under consideration.

V

variable   A letter that represents a number.

Venn diagram   A type of diagram that shows how certain sets are related.

vertex   The point at which two lines, line segments, or rays meet to form an angle.

W

whole number   Any number in the set 0, 1, 2, 3, 4...

X

x-axis   The horizontal axis on a coordinate plane.

x-coordinate   The first number in an ordered pair. Also called the abscissa.

Y

y-axis   The vertical axis on a coordinate plane.

y-coordinate   The second number in an ordered pair. Also called the ordinate.

Important Symbols

<	less than		π	pi
$\leq$	less than or equal to		{ }	set
>	greater than		| |	absolute value
$\geq$	greater than or equal to		$.\overline{n}$	repeating decimal symbol
=	equal to		1/a	the reciprocal of a number
$\neq$	not equal to		%	percent
$\cong$	congruent to		(x,y)	ordered pair
( )	parenthesis		$\perp$	perpendicular
[ ]	brackets		| |	parallel to
{ }	braces		$\angle$	angle
...	and so on		$\in$	element of
• or ×	multiply		$\notin$	not an element of
∞	infinity		$\cap$	intersection
a^n	the nth power of a number		$\cup$	union
$\sqrt{}$	square root		$\subset$	subset of
Ø, { }	the empty set or null set		$\not\subset$	not a subset of
$\therefore$	therefore		$\triangle$	triangle
°	degree			

Multiplication Table

x	2	3	4	5	6	7	8	9	10	11	12
2	4	6	8	10	12	14	16	18	20	22	24
3	6	9	12	15	18	21	24	27	30	33	36
4	8	12	16	20	24	28	32	36	40	44	48
5	10	15	20	25	30	35	40	45	50	55	60
6	12	18	24	30	36	42	48	54	60	66	72
7	14	21	28	35	42	49	56	63	70	77	84
8	16	24	32	40	48	56	64	72	80	88	96
9	18	27	36	45	54	63	72	81	90	99	108
10	20	30	40	50	60	70	80	90	100	110	120
11	22	33	44	55	66	77	88	99	110	121	132
12	24	36	48	60	72	84	96	108	120	132	144

Commonly Used Prime Numbers

2	3	5	7	11	13	17	19	23	29
31	37	41	43	47	53	59	61	67	71
73	79	83	89	97	101	103	107	109	113
127	131	137	139	149	151	157	163	167	173
179	181	191	193	197	199	211	223	227	229
233	239	241	251	257	263	269	271	277	281
283	293	307	311	313	317	331	337	347	349
353	359	367	373	379	383	389	397	401	409
419	421	431	433	439	443	449	457	461	463
467	479	487	491	499	503	509	521	523	541
547	557	563	569	571	577	587	593	599	601
607	613	617	619	631	641	643	647	653	659
661	673	677	683	691	701	709	719	727	733
739	743	751	757	761	769	773	787	797	809
811	821	823	827	829	839	853	857	859	863
877	881	883	887	907	911	919	929	937	941
947	953	967	971	977	983	991	997	1009	1013

Squares and Square Roots

No.	Square	Square Root	No.	Square	Square Root	No.	Square	Square Root
1	1	1.000	51	2,601	7.141	101	10201	10.050
2	4	1.414	52	2,704	7.211	102	10,404	10.100
3	9	1.732	53	2,809	7.280	103	10,609	10.149
4	16	2.000	54	2,916	7.348	104	10,816	10.198
5	25	2.236	55	3,025	7.416	105	11,025	10.247
6	36	2.449	56	3,136	7.483	106	11,236	10.296
7	49	2.646	57	3,249	7.550	107	11,449	10.344
8	64	2.828	58	3,364	7.616	108	11,664	10.392
9	81	3.000	59	3,481	7.681	109	11,881	10.440
10	100	3.162	60	3,600	7.746	110	12,100	10.488
11	121	3.317	61	3,721	7.810	111	12,321	10.536
12	144	3.464	62	3,844	7.874	112	12,544	10.583
13	169	3.606	63	3,969	7.937	113	12,769	10.630
14	196	3.742	64	4,096	8.000	114	12,996	10.677
15	225	3.873	65	4,225	8.062	115	13,225	10.724
16	256	4.000	66	4,356	8.124	116	13,456	10.770
17	289	4.123	67	4,489	8.185	117	13,689	10.817
18	324	4.243	68	4,624	8.246	118	13,924	10.863
19	361	4.359	69	4,761	8.307	119	14,161	10.909
20	400	4.472	70	4,900	8.367	120	14,400	10.954
21	441	4.583	71	5,041	8.426	121	14,641	11.000
22	484	4.690	72	5,184	8.485	122	14,884	11.045
23	529	4.796	73	5,329	8.544	123	15,129	11.091
24	576	4.899	74	5,476	8.602	124	15,376	11.136
25	625	5.000	75	5,625	8.660	125	15,625	11.180
26	676	5.099	76	5,776	8.718	126	15,876	11.225
27	729	5.196	77	5,929	8.775	127	16,129	11.269
28	784	5.292	78	6,084	8.832	128	16,384	11.314
29	841	5.385	79	6,241	8.888	129	16,641	11.358
30	900	5.477	80	6,400	8.944	130	16,900	11.402
31	961	5.568	81	6,561	9.000	131	17,161	11.446
32	1,024	5.657	82	6,724	9.055	132	17,424	11.489
33	1,089	5.745	83	6,889	9.110	133	17,689	11.533
34	1,156	5.831	84	7,056	9.165	134	17,956	11.576
35	1,225	5.916	85	7,225	9.220	135	18,225	11.619
36	1,296	6.000	86	7,396	9.274	136	18,496	11.662
37	1,369	6.083	87	7,569	9.327	137	18,769	11.705
38	1,444	6.164	88	7,744	9.381	138	19,044	11.747
39	1,521	6.245	89	7,921	9.434	139	19,321	11.790
40	1,600	6.325	90	8,100	9.487	140	19,600	11.832
41	1,681	6.403	91	8,281	9.539	141	19,881	11.874
42	1,764	6.481	92	8,464	9.592	142	20,164	11.916
43	1,849	6.557	93	8,649	9.644	143	20,449	11.958
44	1,936	6.633	94	8,836	9.695	144	20,736	12.000
45	2,025	6.708	95	9,025	9.747	145	21,025	12.042
46	2,116	6.782	96	9,216	9.798	146	21,316	12.083
47	2,209	6.856	97	9,409	9.849	147	21,609	12.124
48	2,304	6.928	98	9,604	9.899	148	21,904	12.166
49	2,401	7.000	99	9,801	9.950	149	22,201	12.207
50	2,500	7.071	100	10,000	10.000	150	22,500	12.247

Fraction/Decimal Equivalents

Fraction	Decimal	Fraction	Decimal
$\frac{1}{2}$	0.5	$\frac{5}{10}$	0.5
$\frac{1}{3}$	$0.\overline{3}$	$\frac{6}{10}$	0.6
$\frac{2}{3}$	$0.\overline{6}$	$\frac{7}{10}$	0.7
$\frac{1}{4}$	0.25	$\frac{8}{10}$	0.8
$\frac{2}{4}$	0.5	$\frac{9}{10}$	0.9
$\frac{3}{4}$	0.75	$\frac{1}{16}$	0.0625
$\frac{1}{5}$	0.2	$\frac{2}{16}$	0.125
$\frac{2}{5}$	0.4	$\frac{3}{16}$	0.1875
$\frac{3}{5}$	0.6	$\frac{4}{16}$	0.25
$\frac{4}{5}$	0.8	$\frac{5}{16}$	0.3125
$\frac{1}{8}$	0.125	$\frac{6}{16}$	0.375
$\frac{2}{8}$	0.25	$\frac{7}{16}$	0.4375
$\frac{3}{8}$	0.375	$\frac{8}{16}$	0.5
$\frac{4}{8}$	0.5	$\frac{9}{16}$	0.5625
$\frac{5}{8}$	0.625	$\frac{10}{16}$	0.625
$\frac{6}{8}$	0.75	$\frac{11}{16}$	0.6875
$\frac{7}{8}$	0.875	$\frac{12}{16}$	0.75
$\frac{1}{10}$	0.1	$\frac{13}{16}$	0.8125
$\frac{2}{10}$	0.2	$\frac{14}{16}$	0.875
$\frac{3}{10}$	0.3	$\frac{15}{16}$	0.9375
$\frac{4}{10}$	0.4